Texas Seashells
A FIELD GUIDE

Harte Research Institute for Gulf of Mexico Studies Series,
Sponsored by the Harte Research Institute for Gulf of Mexico Studies,
Texas A&M University–Corpus Christi

Greg Stunz, General Editor
John W. Tunnell Jr., Founding Editor

Texas Seashells
A FIELD GUIDE

JOHN W. TUNNELL JR.,
NOE C. BARRERA,
and FABIO MORETZSOHN

TEXAS A&M UNIVERSITY PRESS
College Station, Texas

Manufactured in China through Martin Book Management.
(∞) This paper meets the requirements
of ANSI/NISO Z39.48-1992
(Permanence of Paper).
Binding materials have been chosen for durability.

LIBRARY OF CONGRESS CATALOGING-IN-PUBLICATION DATA

Tunnell, John W. Jr., author.
  Texas seashells : a field guide / John W. Tunnell Jr., Noe C. Barrera, and
Fabio Moretzsohn. – First edition.
      pages cm—(Harte Research Institute for Gulf of Mexico Studies series)
  Includes bibliographical references and index.
  ISBN 978-1-62349-167-3 (flexbound : alk. paper)—
  ISBN 978-1-62349-196-3 (e-book)
1. Shells—Texas—Gulf Region—Identification.
I. Barrera, Noe C., author. II. Moretzsohn, Fabio, author.
III. Title. IV. Series: Harte Research Institute for Gulf of Mexico Studies series.
  QL415.T4T86 2014
  594.147'709764–dc23
  2014015112

Jean Andrews autographing a copy of
her 1977 book, *Shells and Shores of Texas*,
at the Houston Museum of Natural
Science, October 2006. Photograph
by F. Moretzsohn.

CONTENTS

Field guides have always made me happy because they conjure up good thoughts of being in the field, the best place to really learn about nature and why I got into this field of marine biology in the first place. I have many field guides in my library, and I love thumbing through them, looking at the diversity of marine algae, seashore vegetation, sponges, seashells from various parts of the world, crustaceans, echinoderms, birds, and terrestrial and marine mammals. Field guides are of immense importance to field biologists, as they help them in the first round of identification while at work in the field. Generally, these field guides are written by the experts on a particular group of organisms who have vast knowledge and expertise, and their years, or perhaps decades, of fieldwork and experience with that group are then condensed to a series of good photos or drawings, as well as key characteristics for identification within a particular geographic region.

Jean Andrews was the first to make a field guide to Texas seashells. As is often the case, she first produced a definitive work on Texas seashells in a larger book, *Sea Shells of the Texas Coast* (1971), and then an updated version, *Shells and Shores of Texas* (1977). Subsequently, two small field guides were published from these larger works: *Texas Shells—A Field Guide* (1981) and *A Field Guide to Shells of the Texas Coast* (1991).

In 1999, I decided that it was time to update Jean Andrews's larger books with expanded coverage to deeper-water and tropical species found offshore on Texas coral reefs and banks. Jean allowed us to use some of her book material and format, and I engaged some of my former students and colleagues in the process of the update. The product of that effort is our *Encyclopedia of Texas Seashells* (Tunnell et al. 2010). It includes 900 species found in Texas coastal and marine waters, up from 325 presented in the Andrews books, so its size and weight make it impractical for field use.

Soon after the *Encyclopedia* came out, Shannon Davies, the natural science editor at Texas A&M University Press, said they began to get requests for a field guide version of the big book. So we began the process of scaling back the big book to pocket size. First, we engaged some of our marine biology field colleagues (Larry Hyde, Rick Kalke, and Kim Withers) to help us select the most common species and appropriate ones for the field guide version. Then we had to decide which chapters of the *Encyclopedia* were relevant to the field guide and which were not.

This field guide, therefore, is a presentation of the most common Texas seashells and the necessary information for identification of those species. Although Jean Andrews passed away just two months before our *Encyclopedia* was released, she was the inspiration for both the larger volume and this field guide, and she did have a chance to review the proofs of the big book before it was published and was very happy with it.

I hope this shorter version of the big *Encyclopedia, Texas Seashells: A Field Guide,* will make you happy when you take it with you to Texas bays and beaches to identify the seashells you find on your field trip.

John W. Tunnell Jr.

ACKNOWLEDGMENTS

It takes many people and organizations to compile the necessary information, locate and photograph specimens, and find all the pertinent literature for a volume such as the *Encyclopedia of Texas Seashells*. Museum collections and the people who work with them are acknowledged, particularly for providing assistance with specimens and information: Houston Museum of Natural Sciences—Lisa Rebori, John Wise, Tina Petway, John McAlpin, Bianca Guerrero, and Eydie Rojas; Texas Cooperative Wildlife Collection—Mary Wicksten; Brazosport Museum of Natural Science; and Bailey-Matthews Shell Museum—José H. Leal.

Several shell collectors provided specimens from their personal collections for study, including Roe Davenport, Emilio F. García, Janey Nill Cormier, and Roger Bennett. Many others provided taxonomic assistance with certain species or groups: Gary Rosenberg (Academy of Natural Sciences of Philadelphia); Amélie H. Scheltema (Woods Hole Oceanographic Institution); Daniel Geiger (Santa Barbara Museum of Natural History); James McLean (Natural History Museum of Los Angeles County); Alan Kohn (University of Washington); Emilio F. García (University of Louisiana at Lafayette); Robert Robertson (Academy of Natural Sciences of Philadelphia); Paula Mikkelsen (Paleontological Research Institution); José Leal (Bailey-Matthews Shell Museum); Bill Lyons; and Donna Turgeon. We particularly thank Paula Mikkelsen and Emilio F. García for their excellent reviews and helpful suggestions. Rebekah Thomas (Texas A&M University–Corpus Christi) did a final and thorough review of the entire book, which considerably improved the manuscript. Jochen Gerber (Field Museum of Natural History) and Stephanie A. Clark (Chicago Academy of Sciences) also made useful suggestions.

Several friends from Texas shell clubs shared information, invited us to give presentations about the book, and provided good suggestions, including Leslie Crnkovic, Tina and Frank Petway, Darwin Alder, and Lucy Clampit (Houston Conchology Society); Jean Dickman and Bob Nixon (San Antonio Shell Club); Theresa Stelzig, Alice Pullin, and Teri White (Coastal Bend Shell Club); and Janey Nill Cormier, Wanda and Steven Coker, and Wayne and Patty Humbird (Sea Shell Searchers of Brazoria County). We thank Harry G. Lee of Jacksonville, Florida, Marlo Krisberg of Merritt Island, Florida, and many others at Conch-L for information on Gulf of Mexico mollusks. We also thank Tom Eichhorst, editor of *American Conchologist,* for help publicizing the book.

Roger Bennett (Texas Commission on Environmental Quality) provided some specimens and excellent habitat/field photos, and Russell Hooten (Texas Parks and Wildlife Department) drew several specimens by hand. Personnel from the Center for Coastal Studies and Harte Research Institute for Gulf of Mexico Studies at Texas A&M University provided various kinds of assistance, but we individually thank Gail Sutton, Blenda Bligh, Dixie Smith, Brien Nicolau, and the late Carl Beaver.

Emma Hickerson and G. P. Schmahl (Flower Garden Banks National Marine Sanctuary), Quenton Dokken (Gulf of Mexico Foundation), and Steve Gittings (National Oceanic and Atmospheric Administration) assisted with diving and access to the Flower Garden Banks National Marine Sanctuary. John Woelke and Jeff Janko provided photographic expertise for most macromollusks (those larger than 10 mm), and Tom C. Shirley (Harte Research Institute) made available his laboratory and digital camera that was used for some microphotography. Femorale.com (José and Marcus Coltro) provided photographs of *Aliger gallus,* Emma Hickerson contributed a photo of a live *Nodipecten fragosus,* and Doug Weaver (Harte Research Institute) supplied images of multibeam and side-scan sonar, as well as 3-D models from offshore banks in Texas.

We are especially thankful that a group of individuals and organizations provided funding so that the book could be published in full color: Will Harte; Houston Museum of Natural Science (Lillie and Roy Cullen Endowment Fund); Rotary Club of Corpus Christi (Harvey Weil Trust); Houston Conchology Society; Harley Moody;

J. Oscar Robinson; San Antonio Shell Club; Suncoast Conchologists; Stephen and Nancy Browning; Richard Hardin; Lillian Murray; Jan Roberts; Coastal Bend Shell Club; Brazosport Museum of Natural Science; Sea Shell Searchers of Brazoria County; and North Texas Conchological Society.

Fabio Moretzsohn also would like to remember the late Osmar Domaneschi (Universidade de São Paulo, Brazil) and E. Alison Kay (University of Hawaii at Manoa), his undergraduate and PhD advisers, respectively, both of whom recently passed away, for being such influential mentors and for introducing Fabio to malacology.

We dedicated the *Encyclopedia of Texas Seashells* to our good friend Roe Davenport (March 8, 1939–October 10, 2005). Roe's passion and love for, and detail of work on, Texas seashells inspired us to begin this project. His untimely death strengthened our resolve to finish it. Roe's monthly trips along the Texas coast produced not only a fantastic seashell collection but also a fantastic collection of friends. He was a joy and inspiration to all of us.

The above acknowledgment covers our thanks to people and organizations that helped with the development and publication of the *Encyclopedia of Texas Seashells.* For *Texas Seashells: A Field Guide,* we engaged some of our marine biology field colleagues (Larry Hyde, Rick Kalke, and Kim Withers) to help us select the most common species and appropriate ones for the field guide version. Their knowledge and help in this process are greatly appreciated.

Wes Tunnell also sincerely thanks Will and Pam Harte for providing a quiet cabin at their Caldwell Ranch in the Davis Mountains of far West Texas. This allowed a full and final review of the entire manuscript before submission to TAMU Press without interruption.

Last, but certainly not least, we wish to thank our families for their continuing support during this long production process, especially Kathy Tunnell, Alma B. Barrera, and Heather Moretzsohn.

INTRODUCTION

Seashells have intrigued humans for eons, even before recorded history, as noted in prehistoric drawings and artifacts found all over the world. Seashells have been used as tools, containers, objects of adornment, fetishes, and even currency in early cultures. Chapter 1 in our *Encyclopedia of Texas Seashells* (Tunnell et al. 2010) is dedicated to the early prehistoric and historic use of seashells along the Texas coast. Today's use of mollusks ranges from important and highly valued food sources of shellfish, such as oysters, mussels, and clams, to highly prized jewelry items like pearls and cameos, to whole or partial shell ornaments decorating homes.

Seashells found on the shores of our Texas bays and Gulf beaches are the exoskeleton or outer, protective shell covering of a mollusk. Mollusks are a category, or phylum, of invertebrates found within the animal kingdom. The phylum is subdivided into 7 classes of different kinds of mollusks based on features of their body and shell shape and biology. Five of these 7 classes of mollusks are found on the Texas coast, but only 2 of the 5 are readily recognizable to the casual observer as seashells: the bivalves, or clams, oysters, and mussels; and the gastropods, or snails, conchs, and whelks. The purpose of this field guide is to assist the beachgoer, student, or scientist in identifying the more common species of seashells found on the Texas coastline. The book is intended to be taken to the shore and used in the field.

This field guide is an abridged version of our larger *Encyclopedia of Texas Seashells* published by Texas A&M University Press in 2010. The field guide covers 300 of the better-known or more common species out of the 900 species in the *Encyclopedia,* and it leaves out a lot of details not needed for the field identification of species. In this guide, after two short chapters regarding shell collecting and identifying seashells, we jump right into the species accounts. We illustrate each species on the same page with its description to allow viewing the picture while reading the text. All photographs are in full color, and the adjacent text gives the current scientific name and common name, if there is such (many very small species do not have common names), distribution, size, description, remarks, and, in some cases, synonyms, or former names used for the species.

The sequence of classes and families from front to back in this field guide is from the most primitive to the most advanced. Class and family descriptions help the reader learn more about the larger groups in which the species belongs, and the species are listed in alphabetical order under the families.

Classes of mollusks or seashells within the field guide follow this sequence:

- Polyplacophora (chitons)
- Gastropoda (snails, conchs, and whelks)
- Cephalopoda (squids and octopuses)
- Bivalvia (oysters, mussels, and clams)
- Scaphopoda (tuskshells)

A glossary is provided at the end for technical terms that are not defined within the text.

For further information on all topics and species within this field guide, we refer the reader/user to the *Encyclopedia of Texas Seashells.*

Texas Seashells

A FIELD GUIDE

CHAPTER ONE

Collecting Seashells

Introduction

Many people start to collect shells by picking up seashells on a visit to the beach or perhaps land snails in their own backyard. Whether they immediately become interested in collecting shells or wait many years for another opportunity to visit the shore, they often want more shells in their collection. There are basically three ways to build a shell collection: (1) by collecting the shells yourself, (2) by trading them with other collectors, and (3) by purchasing them from dealers. The purpose of this field guide is to facilitate the first option, collecting your own shells and then identifying them yourself. Collecting seashells can be very enjoyable, and identification of your collection can be quite gratifying. We offer these hints to assist you in your collecting and identifying adventures.

Regulations and the "Sheller's Creed"

Before you start on a shell-collecting trip, be sure to check the local laws. Many places regulate and limit the number and/or species that can be live collected; the area may be closed for any live collecting on a seasonal basis or permanently; and you may need a permit. You need a valid fishing license with a freshwater or saltwater stamp to take live mollusks in the public waters of Texas (TPWD 2006: 22); however, you do not need one to take empty or dead shells.

At the time of this writing (April 2013), the only place in Texas that currently has laws limiting collection of live shells is South Padre Island, where there is a combined daily limit of 15 live univalve shells (all species), including no more than 2 each of the following species:

lightning whelk (*Busycon pulleyi*); pearwhelk (*Busycotypus spiratus*); horse conch (*Triplofusus giganteus*); Florida fighting conch (*Strombus alatus*); banded tulip (*Fasciolaria tulipa*); and Florida rocksnail (*Stramonita haemastoma*) (Blankinship et al. 2005; TPWD 2006: 28). Additionally, there is an annual "no-collection" period, from November 1 to April 30, which prevents the taking of any live or dead mollusks or their shells (including those with hermit crabs), starfish or sea urchins within the area bounded by "the bay and pass sides of South Padre Island, and from the east end of the north jetty at Brazos Santiago Pass to the west end of West Marisol Drive in the town of South Padre Island, out 1000 yards [914 m, or 3000 ft] from the mean-tide line, and bounded by the centerline of Brazos Santiago Pass" (TPWD 2006: 28).

Some areas are protected for conservation, such as the Flower Garden Banks National Marine Sanctuary (FGBNMS), where collection can be done only by persons granted a scientific permit. Visit the FGBNMS website for more information: http://flowergarden.noaa .gov.

The Hawaiian Malacological Society (HMS) proposed "A Sheller's Creed" and first published it in *Hawaiian Shell News* in August 1973 to guide its members when collecting live shells. Many shell clubs and collectors around the world have adopted this short code of conduct, and as marine biologists we encourage our readers also to adopt it. It is reproduced here by kind permission from HMS:

A Sheller's Creed

The wildlife and natural resources of this world have been entrusted to me for protection and preservation. Whether I wish it or not, I must account to the future for my handling of this wealth today. If I collect shells, I will do it conservatively, recognizing that destruction of the marine habitat, by whatever means, is the true enemy of the sea and its creatures.

Four rules to shell by:

1. Leave the live coral heads alone! Look in the rubble, under the slabs, in the sand, and among the loose chunks.

2. Put rocks and corals back in place, the way you found them, even in deep water. Many things live under them even if you do not recognize them. Continued exposure to light, predators, and current will kill many of them.

3. Be alert for shell eggs and protect them. They have a slim chance of survival, at best. Don't take the mollusk that is guarding them. Avoid disturbing breeding groups.

4. Collect only what you really need. Take time to examine your finds and leave them to grow and breed if they do not really meet your needs.

A couple of rules could be added to the Sheller's Creed:

5. Do not collect heavily damaged live shells or ones with break-repair scars because they are not valuable for trading (although they might be important for scientists). Leave them behind, and the next season they may produce healthy specimens. And right before you leave the shelling grounds, do a quick inspection of your catch and return the damaged and unwanted shells to the ocean.

6. Do not litter or pollute our waters.

Collecting Seashells

Most of the Texas coast has sandy beaches, marshes, and bays, basically all composed of soft substrate (for information on habitats, see Hicks 2010; Britton and Morton 1989), but there are also jetties, oyster reefs, and a few other hard substrates. Equipment needed for collecting mollusks varies with the species you want to collect or how much you want to specialize. The basic equipment can be as simple as some vials, a kitchen strainer, a spade, a knife, sharp eyes, and lots of patience! As you advance in your collecting skills, you can add a few more pieces of equipment, such as screens, dips and nets, shovels, a hammer, a water pump, a glass-bottomed bucket, snorkeling gear, a light (for night collecting), a loupe (small magnifying lens), a wire or probing tool, forceps, and tweezers. Other useful tools include maps, a GPS (global positioning system) receiver, camera (some recent point-and-shoot digital camera models are waterproof to shallow depths), notebooks, field guides, weather-appropriate clothing, towels, fresh clothes (for after you finish collecting), tide table, insect repellent, sunscreen, and first-aid kit (see Sturm et al. 2006 for equipment and how to use it, and field techniques).

Probably most shell collecting in Texas is done at sandy beaches or while wading in bays, since these habitats dominate the coast. Some people snorkel along the coast or scuba dive at the offshore oil rigs or at the Flower Garden Banks National Marine Sanctuary. Many of the shells illustrated in Tunnell et al. (2010) came from depths beyond the normal recreational diving depths; they have been collected by dredging or through remotely operated vehicles (ROV) and submarines. We used the same photographs in this book, with a few exceptions of photos of better shells, but most of the species herein are common and occur in shallow, coastal waters and can be collected while wading or walking on the beach.

Small and miniature shells, also known as micromollusks, are difficult to pick up in the field. People interested in collecting micros usually collect beach drift, shell grit, or sand samples at the beach, or when snorkeling or scuba diving, and bring the samples home or to the laboratory. Samples can be placed on trays, air-dried, and sorted by size using a series of sieves with different mesh sizes, and then studied under a stereomicroscope.

Live micromollusks can be collected through several specialized techniques, such as rinsing seaweeds and brushing rocks into a plastic bag or pillowcase while scuba diving. Live samples can be narcotized with different chemicals (e.g., magnesium salts) or low temperature to be studied under a microscope or to relax the animals for fixation. For more information about collection, curation, and other techniques to study micromollusks, see Geiger et al. (2007).

Trading Shells

Trading shells can be rewarding, and it is a good way to replace duplicate shells, or publications about shells and collecting, with something new. Before you start trading shells, you should learn about the trader's reputation. Make sure the trader provides good-quality shells with reliable collecting data and that the trade is fair in terms of number, quality, and rarity of the shells. The main point is that both parties must be satisfied with the trade after it is completed. Therefore, you should start with modest trades or purchases (if buying from a dealer) until you can be sure that the trader can be trusted. As most trades are usually for self-collected specimens, collection data from a trader may be more complete and reliable than that from a dealer.

Shell Grading

In 1973, Elmer Leehman and Stu Lilico proposed what is known as the Hawaiian Malacological Society International Shell Grading Standard

(HMS-ISGS), a system to help shell collectors and dealers to objectively grade shells. The system, with some slight modifications, has been widely adopted by the shelling community and basically has the following categories:

- Gem (G)—A fully mature specimen, without any visible blemishes, perfect protoconch (larval shell), no broken spines and lips, rich color, live collected.
- Fine (F)—An adult shell with only minor flaws, with color and gloss, minor growth marks or a small chip on the lip, 1 or 2 broken spines, but no repairs such as filed lips.
- Good (Gd)—An acceptable shell, despite some defects such as broken spines, worn spire, and growth marks but still a good representative of the species. Not necessarily live taken. The nature and degree of repairs should be disclosed.
- Commercial (C) or Poor (P)—Dead or beach collected, with growth marks, chipped lips, or broken spines. No data or operculum. Should not be offered as collectors' specimens.

Following are other abbreviations and symbols often used with shell grading to describe the condition of the shell:

- W/O = with operculum (a plate that plugs the aperture of a gastropod)
- W/P = with periostracum (fibrous coating that covers the shell)
- F/D = full data (provenance, habitat, date, and original collector)
- B/D = basic data (less than full data)
- JUV = juvenile or subadult
- + (plus) or − (minus)—used in conjunction with shell grades to denote borderline cases (e.g., G−, F+, etc.)
- P.O.R. = price on request (used sometimes for high-end specimens)

Maintaining a Shell Collection

The Mollusks—A Guide to Their Study, Collection, and Preservation (Sturm et al. 2006), published by the American Malacological Society (AMS), is a "must have" for any collector. This book, the result of the Malacology Curation Workshop at the AMS annual meeting in Pittsburgh in 1999, covers a wide range of topics of interest to the shell collector, including how to organize a shell collection, collection techniques, curation procedures and supplies, photography, an introduction to all molluscan classes, a primer on taxonomy, phylogenetics, conservation, and other related topics.

Shell Clubs

Shell clubs are groups of people interested in shell collecting, nature, the outdoors, fossils, and other related topics who meet on a regular basis to "talk shells." Shell clubs are a good place to meet other people with similar interests, make contacts with other collectors, and trade specimens. Typically, shell clubs meet on a regular basis, usually once a month at a museum, school, church, public building, or other available place, to get updates on the members' shelling activities or news. Often there is a shell-related presentation by one of the members or a guest speaker. Members and their guests often bring shells that they cannot identify or a new acquisition that they want to show their friends. It is also common to see people trading shells and/or shell dealers offering local and worldwide shells.

Some clubs organize and sponsor field trips and other events such as shell shows that are open to the public. Such shows may have competitions for best displays on a certain theme, as well as for the best, rarest, largest, smallest, or most abnormal ("freak") self-collected shell or for shells with historical value or most interesting story. Shell shows tend to attract a lot of visitors and become an important source of new members.

There seems to be an unfortunate general trend of shrinking memberships in shell clubs, although a few clubs report growing membership. Some of the reasons for this trend include increasing restrictions on live shell collecting in many places; kids more interested in video games, TV, and computers than playing outdoors in nature; kids being taught at school that it is bad for the environment to take anything, especially live, from the ocean; habitat changes, pollution, and over-collecting that have led to a drastic reduction in the numbers of shells in some places; life has become too busy for shells. The result is that many clubs are experiencing an aging membership, with little, if any, recruitment of young members. The HMS, for example, had almost 1700 members worldwide at its peak in the 1970s, but membership has steadily declined so that it is now a fraction of that figure. Some clubs have recently closed because of lack of interest. On the other hand,

Internet search engines receive millions of queries about shells, suggesting that there is still a large interest in shells.

Currently there are 6 active shell clubs in Texas:

- Coastal Bend Shell Club—Corpus Christi (no website)
- Houston Conchology Society—Houston (http://www.houstonshellclub.com/)
- North Texas Conchological Society—Dallas-Fort Worth (http://outdoorplace.org/shells/)
- Port Isabel /South Padre Island Shell Club—Port Isabel (no website)
- San Antonio Shell Club—San Antonio (http://www.sashellclub.org/index.html)
- Sea Shell Searchers of Brazoria County—Clute (http://bcfas.org/museum/SSSBC/SSSBC.html)

Internet Resources

There are several online forums that discuss shells, such as these:

- Biodiversity of the Gulf of Mexico Database (BioGoMx)—A database created and maintained by the Harte Research Institute that lists all living species from the Gulf of Mexico. Available at http://www.gulfbase.org/biogomx/.
- Conch-L—A forum for shell collectors, with many friendly and knowledgeable people willing to help fellow shellers. Its archive can be seen at http://www.listserv.uga.edu/archives/conch-1.html.
- Let's Talk Seashells—Created and maintained by Marlo Krisberg, it is a forum for seashell collectors that includes help with identification, tutorials, great photos, and other tools. Visit it at http://www.letstalkseashells.com/.
- Malacolog—Created and maintained by Gary Rosenberg, Academy of Natural Sciences of Drexel University, it is a taxonomic database on western Atlantic marine mollusks. Search it at http://www.malacolog.org/.
- Mollusca-L—An e-mail listserv primarily dedicated to malacologists and students. For more information and subscription, see http://www.ucmp.berkeley.edu/mologis/mollusca.html.

- World Register of Marine Species (WoRMS)—A taxonomic
  database on marine organisms, maintained by experts. Available
  at http://www.marinespecies.org/.

National Organizations

- The American Malacological Society (AMS)—This professional
  association of malacologists publishes a scientific journal
  (*American Malacological Bulletin,* of which Fabio Moretzsohn
  is the managing editor) and holds an annual meeting. Website
  includes links to other associations and websites: http://www
  .malacological.org/.
- Conchologists of America (COA)—This society for shell
  enthusiasts publishes the quarterly magazine *American
  Conchologist;* holds an annual convention; includes links on
  its website to worldwide shell events, meetings, shell clubs, and
  other shell-related sites: http://www.conchologistsofamerica.org/

Conchology versus Malacology

The general perception is that conchology is the study of the shell
and is mostly the realm of amateurs, while malacology is the study of
the organism (including the shell) done by scientists. However, some
people, like Lovell Reeve in the nineteenth century, make the case that
there should be no distinction between the two terms, since the shell
is part of the animal, but prefer the term *conchology.* Therefore, both
terms are sometimes used interchangeably.

CHAPTER TWO

Seashell Characteristics

Seashells belong to the phylum Mollusca, one of the most diverse groups of animals, second only to the arthropods. Estimates of the diversity of living mollusks vary from 52,525 to more than 200,000 species; most authors accept an intermediary figure. However, recent studies suggest that perhaps as many as half of the mollusks remain undescribed. Additionally, there are some 35,000 to 70,000 fossil species.

The majority of mollusks live in the oceans, occupying all habitats, from the intertidal and splash zones to the bottoms of marine trenches. The deepest-living mollusk has been reported from 8595 m (28,200 ft) deep in the Cayman Trench. About 7000 species of bivalves and gastropods live in brackish or fresh water, and some 24,000 gastropods live on land in nearly all habitats, including forests, deserts, and even mountaintops. Mollusks have a long fossil history dating back to the Paleozoic era (550 million years ago), and thanks to their hard shells, mollusks have one of the most complete fossil records.

As currently recognized, the phylum Mollusca includes 7 classes of living species: Aplacophora, Polyplacophora, Monoplacophora, Gastropoda, Cephalopoda, Bivalvia, and Scaphopoda. See table 2.1 for a summary of the number of species per class. Texas has representatives of all classes but Monoplacophora. Most species of aplacophorans occur in deep water and therefore are not represented in this field guide; refer to *Encyclopedia of Texas Seashells* (Tunnell et al. 2010) for more information.

Mollusks display a remarkable diversity in body shape but share a common body plan: they are bilaterally symmetrical (mirror image),

TABLE 2.1. Summary of the species numbers per class

Class	Living species (estimates)	Species in the Gulf of Mexico	Species in Texas	Species in this book
Aplacophora	120	10	4	0
Monoplacophora	25	0	0	0
Polyplacophora	900	41	7	2
Gastropoda	70,000	1742	593	156
Cephalopoda	900	93	10	4
Bivalvia	20,000	528	276	137
Scaphopoda	900	41	10	1
Total	~93,000	2455	900	300

Source: Living species (based on Brusca and Brusca 2003), species listed in *Biota of the Gulf of Mexico* (Felder and Camp 2009), species in Texas (Tunnell et al. 2010), and species illustrated in this book.

unsegmented, lower invertebrates. The main body cavity is their open circulatory system, which has a dorsal heart in a pericardial chamber. The normal body cavity is reduced to small spaces around the gonads, kidneys, and part of the intestine. The body is typically organized in a head-foot–visceral mass morphology, with sensory organs in the head. The foot is large and muscular and, in most cases, has a creeping sole. The mouth area bears a radula, a tonguelike organ with teeth on it, used in feeding, except in bivalves, which lack the radula.

Molluscan taxonomy, as in other groups, is in flux, and as a result, classification is often changing. Among the main changes is a departure from the "traditional" higher classification of gastropods, for example, dividing prosobranchs into Archaeogastropoda, Mesogastropoda, and Neogastropoda, to a cladistic or phylogenetic classification, based on evolutionary history. These changes were needed because new research into evolutionary history showed that some of the former groups incorrectly grouped mollusks that have similar shell morphology but are not closely related.

The taxonomy used in this book follows that adopted in Tunnell et al. (2010), which is an updated taxonomy and therefore differs considerably from that used in *Shells and Shores of Texas* (Andrews 1977). Here we provide some of the more common synonyms; refer to Tunnell et al. (2010) for more information about taxonomic changes.

Although there are only a few documented cases of marine mollusks becoming extinct in modern times, there are some conservation issues. In addition, many mollusks are harvested commercially, and these have traditionally been managed as fisheries; only recently have they been considered for conservation. Two examples of troubled molluscan fisheries in the Gulf of Mexico are the eastern oyster (*Crassostrea virginica*) and the queen conch (*Eustrombus gigas*). The former is not only economically but also ecologically important, and the status of this fishery has been a concern for a long time. There are currently several oyster reef restoration projects attempting to bring oysters back to some parts of the Texas coast where they once were abundant. The queen conch fishery has collapsed in many places, and now the species is protected in Florida and elsewhere in the Caribbean, but it is still harvested in others, such as the Bahamas. Queen conchs occur only in offshore banks in Texas and are not harvested.

Increasing human population along the Gulf of Mexico coast, habitat loss, declining fisheries in general, pollution, hypoxia, climate change, and the recent Deepwater Horizon oil spill (2010) have placed increased pressure on natural resources and, as a consequence, brought about increased regulation. Besides commercial fisheries, mollusks support important recreational fisheries in the Gulf of Mexico, such as clams, oysters, scallops, conchs, whelks, and other gastropods.

Important books on molluscan biology include those by Fretter and Graham (1962), Yonge and Thompson (1976), and Brusca and Brusca (2003). Beesley et al. (1998) compiled an invaluable reference on the biology and systematics of mollusks; although focused on the Australian fauna, the review and information on each family and higher categories are useful globally. Two recent books on bivalves are important references: Mikkelsen and Bieler (2008) on bivalves of southern Florida, and Valentich-Scott and Coan (2012) on West Coast bivalves. In addition to books on Texas seashells (Andrews 1977; Tunnell et al. 2010), other useful works that may complement this book and help with identification of Texas seashells include Abbott (1974), Lee (2009), and Redfern (2013).

Features of Each Class

POLYPLACOPHORA

Polyplacophorans, or chitons, are dorso-ventrally flattened mollusks with a shell divided into 8 plates or valves and a broad, muscular foot

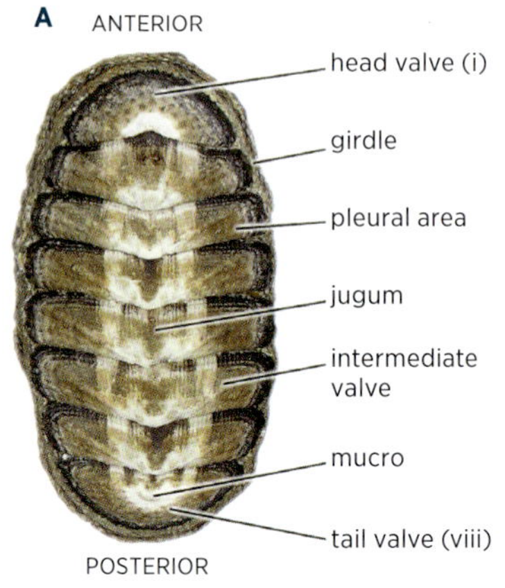

FIGURE 2.1. Polyplacophora (chiton) features of *Chiton granulosus* Frembly, 1827 (from Chile). A: dorsal view; B: ventral view; C: anterior view showing girdle and nodules on valves; D: *Acantochitona lineata* Lyons, 1988: intermediary valve showing apophyses. Photographs A, B, and C by F. Moretzsohn; D by N. Barrera.

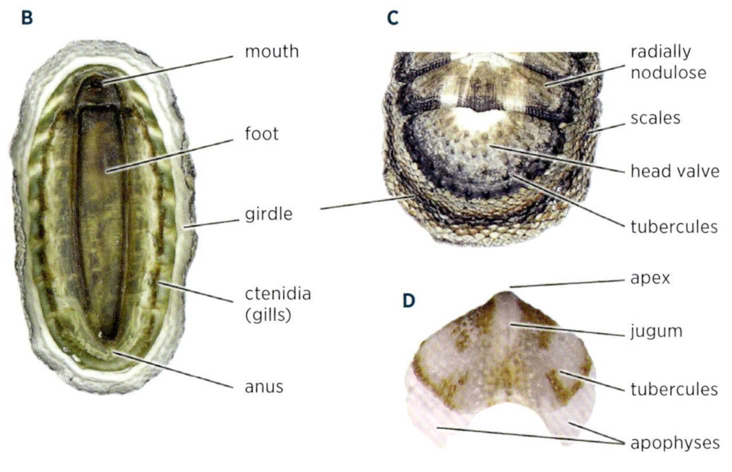

(fig. 2.1). Chitons are exclusively marine, occurring worldwide, with the highest diversity in the warm temperate zones of the world. They live on hard substrates, including empty seashells (a few live in coarse sand), from the splash zone down to the deep sea. They range in size from a few millimeters to more than 300 mm (12 in) in length. The largest species is the giant Pacific chiton, *Cryptochiton stelleri* (von Middendorff, 1847), found from California to Alaska.

Chitons lack true eyes but have sensory organs embedded in their shell that may react to light and touch. Most species run away from light, hiding under rocks and in crevices during the day and becoming active at night. Some species move within a certain area or path, and experimental studies have demonstrated a "homing" behavior, similar to that of limpets. When removed from the substrate, chitons tend to roll up ventrally like pillbugs. Most chitons are herbivores and actively scrape macroalgae and biofilm off rocky substrates, an action that erodes the substrate over time; others feed on sponges or other animals, such as small crustaceans.

There are more than 900 recognized species of chitons in the world, of which 41 occur in the Gulf of Mexico and 7 in Texas; the 2 most common species are discussed here.

GASTROPODA

The gastropod shell is a hollow cone coiled around a central axis, the columella. The base of the cone is the aperture, where the body (head and foot) of the gastropod withdraws to protect itself from predators. Each complete coil of the cone is called a whorl, with the outermost or last one known as the body whorl. The other whorls make up the spire, which includes the apex (fig. 2.2). The oldest part of the shell (the protoconch, the larval shell of the veliger), sits at the top of the apex, although it is fragile and often missing. The shell grows by the mantle's secreting calcium carbonate in small increments on the outer lip of the shell. The shell is reduced or absent in some gastropods (e.g., nudibranchs lack a shell).

Important shell features used in taxonomy include shape, size, color, texture, and aperture shape. However, variation within a species can be great, thus making shell characters nearly meaningless in some groups. Other taxonomic characters include the radula, anatomical features of the stomach, gills, and other organs. Molecular characters are increasingly used more often and complement anatomical and shell characters.

The animal is permanently attached to the columella by the strong columellar muscle, which withdraws the animal into the shell when contracted. Most gastropod shells are dextral, or right-handed: with the shell spire pointing up and the aperture facing you, the aperture is to the right of the columella, where you can insert your hand into the aperture in larger shells and your fingers curl around the columella. Some gastropods have a sinistral, or left-handed, shell (e.g., the lightning whelk). Rare mutants may occur in all species, causing the gastropod to coil in the opposite way.

Gastropods vary in shell size from less than 1 mm (¹⁄₂₅ in) (e.g., minuscule ammonicera, *Ammonicera minortalis*) to over 760 mm

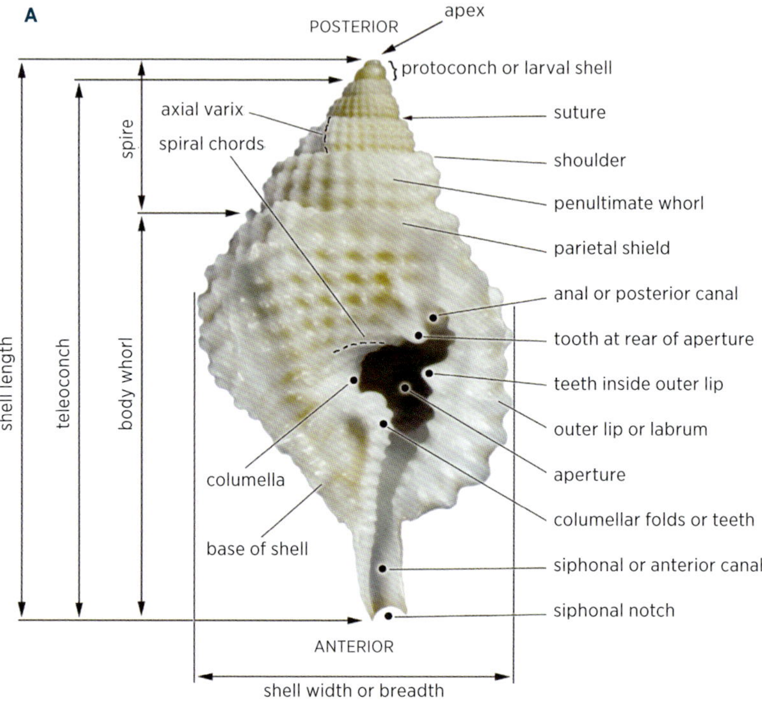

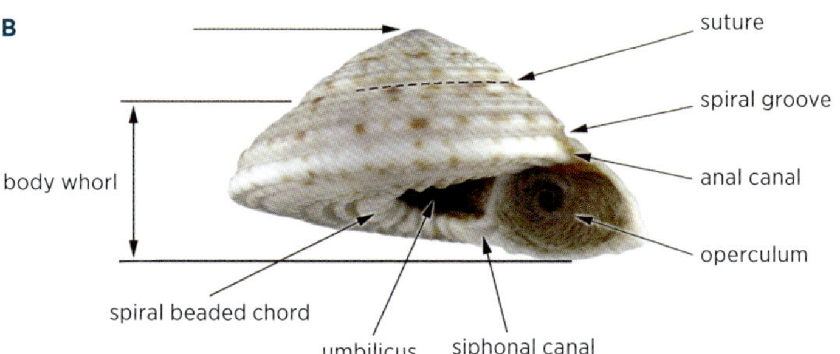

FIGURE 2.2. Gastropod shell features. A: *Distorsio clathrata*: some features on a typical gastropod shell; B: *Architectonica nobilis*: an umbilicate gastropod shell; note the operculum. Photographs by F. Moretzsohn.

(30 in) in length (Australian trumpet, *Syrinx aruanus*). The largest Atlantic gastropod is the horse conch (*Triplofusus giganteus*), which grows to about 600 mm (24 in) in length. The Texas State Shell, the lighting whelk (*Busycon pulleyi* Hollister, 1958), is common and one of the largest shells on the Texas coast.

Adult shells less than 10 mm (⅖ in) in length are considered micromollusks (some authors define them as 5 mm, or ⅕ in, or less). A large proportion of marine gastropods have small to minute shells, and many new species are still being discovered even in well-studied waters. Barrera (2001) studied the micromolluscan fauna of the Flower Garden Banks and found over 100 new records for Texas, including some new records for the Gulf of Mexico; a few are likely species new to science and await a formal description.

The class Gastropoda is the largest in the phylum Mollusca, with an estimated 70,000 living species and some 15,000 fossil species. The Gulf of Mexico is home to at least 1742 species of marine gastropods, of which 593 occur in Texas; 156 species are included in this book.

CEPHALOPODA

This class includes several familiar mollusks, such as the octopus, squid, cuttlefish, and nautilus. Cephalopods are the most specialized group of mollusks, showing little external resemblance to other mollusks. They include the fastest, largest, and most intelligent of all invertebrates. Cephalopods are adapted for locomotion by jet propulsion; they contract the mantle wall and pump water through a muscular funnel (the siphon), making some squids the fastest of all aquatic invertebrates. Their morphology is very different from that of other mollusks (fig. 2.3): they have a ring of muscular appendages (arms or tentacles) used to capture prey or manipulate objects; they possess complex eyes, nervous systems, and behaviors; their parrotlike beak and poison gland are used to kill and tear apart their prey; and they exhibit other features related to a high metabolism and active life. However, they also have molluscan structures such as a radula and a calcareous shell (although in most species it is internal, reduced, or lost, as in octopods).

Cephalopods range in size from 20 mm (⅘ in) to about 15 m (49 ft) in total length; most, however, are between 60 and 700 mm (2 ⅖ and 27½ in). The giant squid (*Architeuthis dux* Steenstrup, 1857), from the North Atlantic (also found in the Gulf of Mexico), is an elusive species and the largest of the invertebrates. Partially digested specimens have

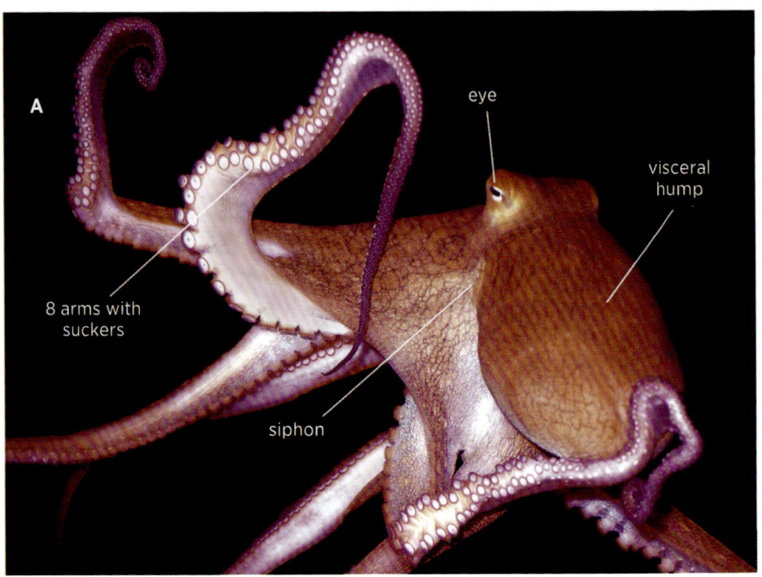

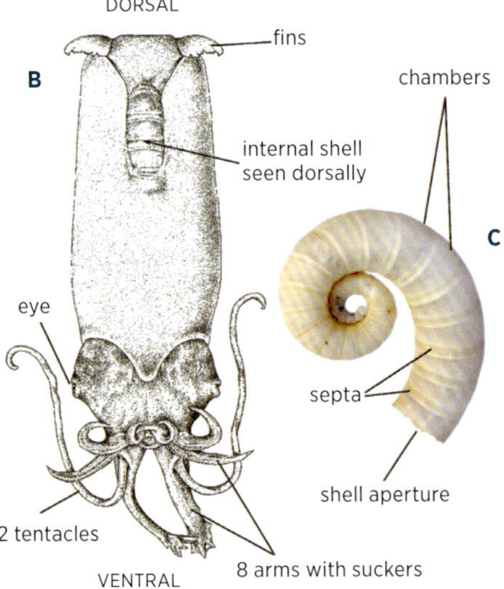

FIGURE 2.3. Cephalopod features. A: *Octopus vulgaris*; B, C: *Spirula spirula*; B: sketch of animal with internal shell showing dorsally; C: internal shell. Photographs A and C by F. Moretzsohn; B from Vecchione (2002).

been found in sperm whale stomachs and sometimes washed ashore. Only a few specimens have been seen or videotaped alive (Roper and Shea 2013). Sailors have reported and greatly exaggerated mythological ocean monsters and legends like the *kraken* (Nordic for *octopus*) since the sixteenth century.

The two main groups of living cephalopods are the superorders Octopodiformes (octopods, with 8 arms, usually with a globular-shaped body, adapted for benthic life) and the Decapodiformes (squids and cuttlefishes, with 8 arms and 2 longer tentacles, for a total of 10 appendages; most species are pelagic, i.e., live in the water column). Squids have a reduced, thin, internal chitinous shell called a gladius or pen, while cuttlefishes have an internal calcareous shell with chambers, called cuttlebone; cuttlefish are not found in the Gulf of Mexico. In addition, the subclass Nautiloidea, among the oldest cephalopod groups, dating back to the late Cambrian (about 505 million years ago), includes many fossil species but only a few living species of *Nautilus,* from the Indo-Pacific region. They are the only living cephalopods with an external shell. The "shell" of the paper nautilus (*Argonauta,* an octopod genus) is not a true molluscan shell but a thin, calcareous brooding chamber secreted by the female.

Cephalopods are active predators of crabs, fishes, other cephalopods, or in the case of octopods, other mollusks such as clams or gastropods. In turn, they are a major food item for many whales, seals, birds, fishes, and other cephalopods. They are commercially important and fished heavily in some countries; over 3 million metric tons are caught every year worldwide.

Cephalopods are exclusively marine. They live in all major marine habitats and occur worldwide, from the intertidal zone to depths exceeding 7000 m (23,000 ft). There are about 900 Recent species worldwide (and over 10,000 fossil species); 93 species live in the Gulf of Mexico and 10 in Texas; 4 are included in this book.

BIVALVIA

The Bivalvia includes familiar mollusks such as clams, mussels, scallops, and oysters. Bivalves are usually bilaterally symmetrical (i.e., mirror images), laterally compressed, and with an external calcareous shell with 2 valves (left and right). The valves are hinged dorsally, connected by an elastic ligament, and held together by adductor muscles (the parts of scallops that we eat) attached to the inner surfaces of the valves. The valves open by the elastic action of the ligament and close by contraction of the adductor muscles.

Several characters are used in bivalve classification, including structure of the gills; size and position of adductor muscles; the degree of fusion of the mantle edges and pallial scars (impression of the mantle) on the inner surface of the shell; mode of life (e.g., burrowing, boring, attaching with a byssus [threads used to anchor the shell], or cementing to a hard substrate); as well as some shell characters, including the hinge dentition and ligament and external surface sculpture of the valves (fig. 2.4). Pallial lines are particularly important in the identification of species that have similar shells, such as some Tellinidae, Lucinidae, and Veneridae; therefore, the pallial line is outlined on some shells in photographs in the species descriptions.

Bivalves range in size from about 1 mm (1/25 in) in length in some condylocardiids, to over 1.5 m (5 ft). The giant clam, *Tridacna gigas* (Linnaeus, 1758), of the tropical Indo-Pacific can weigh more than 330 kg (730 lb). Shell shape varies greatly in bivalves, and the shell can contain spines (e.g., *Spondylus*) or radial ribs (e.g., *Cardita*); can have concentric ridges (e.g., *Dosinia*), radial ribs and nodules (e.g., *Trachycardium*), divaricated lines (e.g., *Nemocardium*), cancellate sculpture (e.g., *Codakia*); or have a combination of different shell sculptures. The shells can be informative about the habitat where the bivalve lives, as well as document environmental conditions and changes over time. Since the shell grows by small, periodic increments, a detailed count of the growth lines can infer the shell's age (like tree rings), and changes in its microchemistry can be used as a record of environmental changes. As some bivalves have been shown to be among the longest-living animals, they can be used to document environmental changes over long periods of time. Recently, an ocean quahog, *Arctica islandica* (Linnaeus, 1767), from Iceland was determined to be more than 405 years old. Most clams, however, probably have a life span of a few years.

Bivalve shells grow by secretion of calcium carbonate by the middle fold of the mantle at the margin of the valves. The oldest part of the shell is the beak (umbones or prodissoconch) near the hinge, and it corresponds to the larval shell. The exterior of the shell is covered by the periostracum, an organic layer that can be thick in Arcidae and Mytilidae, or thin, as in oysters.

Some bivalves have great economic importance because of four factors: (1) They are used as food: clams, oysters, mussels, and scallops

FIGURE 2.4. Bivalve features of *Mercenaria texana*. A: right valve in external view; B: dorsal view; C: left valve in internal view. Photographs by F. Moretzsohn.

account for nearly 6 million tons of catch worldwide. (2) Pearls and mother-of-pearl are valuable. These come mostly from cultured pearl oysters (*Pinctada*) but also from freshwater mussels, which are used for buttons, jewelry, and so on. (3) Biodeterioration causes economic loss, which is difficult to valuate. Shipworms (Teredinidae) bore into water-logged wooden structures and cause great destruction to boats, piers, and other wooden structures; lithophagines (Mytilidae: Lithophaginae) bore into coral and limestone. Some bivalves, such as the invasive mussels (*Perna perna* and *P. viridis*), grow in large densities, fouling rocks, piers, and other hard surfaces (see Tunnell et al. 2010: 318). The freshwater *Corbicula* and *Mytilopsis* clams clog intake pipes in water treatment plants, power plants, and so on. (4) Some suspension-feeding bivalves are used as bioindicators of water quality to monitor bacterial and heavy-metal pollutants.

Most bivalves are free-living and bury in soft sediments (e.g., *Mercenaria*, *Donax*), whereas mussels (Mytilidae) and others produce a byssus and live attached to rocks, shells, or other hard substrates. Oysters and other sessile bivalves cement one of their valves onto hard substrates. Some scallops can actively swim long enough to avoid their starfish predators; the scallops rapidly open and close their valves, causing the shell to move somewhat erratically in the water column.

There are about 20,000 species of bivalves currently recognized in the world, with 528 species occurring in the Gulf of Mexico and 276 in Texas; 137 species are included in this book.

Members of the class Scaphopoda, known as tuskshells, are benthic marine mollusks (a few may be found in estuaries), living buried in soft sediments from soft mud to coarse shell gravel. Although some species occur in the intertidal zone, most live offshore in the deeper waters of the continental shelf and down to depths of 7000 m (23,000 ft). Therefore, scaphopods are rarely mentioned in the literature except to note their presence, and even their distribution is not well documented. In contrast with other molluscan groups, scaphopods are relatively poorly known, especially their anatomy and biology, despite a rich nineteenth-century literature on anatomy mainly on two northeastern Atlantic species, *Antalis entalis* (Linnaeus, 1758) and *A. vulgaris* (da Costa, 1778). Both species have been used as model organisms in embryological studies. Classic monographs from 1897–1898 and 1920 are still considered landmark references for scaphopods.

Scaphopods have a simple, single-valved shell that is a hollowed, tusklike tube open at both ends. It is usually curved and widest at the anterior end (aperture) and narrows toward the posterior end. In the family Gadilidae (see Tunnell et al. 2010: 392), the shell is fattest at about the middle or third of the length of the shell. The shell can be smooth or ornamented, with longitudinal ribs or annular rings. The apex, often notched, is periodically cast off to increase its diameter as the shell grows. The pattern of shell sculpture, shell shape, and apex notch is important in species identification. Shell size varies from about 2 mm (1/12 in) to over 150 mm (6 in) in length, although most reach only about 50 mm (2 in).

Scaphopods are benthic microcarnivores, feeding mostly on foraminifera but also on bivalve spat or detritus. The muscular foot and rudimentary head extend from the aperture. The foot, used to burrow into the soft sediment, is also involved in feeding; it has clubbed contractile tentacles with adhesive tips that capture and move food particles by ciliary gliding.

Currently, scaphopods do not have an economic value, but Pacific Northwest Native Americans once used them extensively as currency called wampum. Wampum consisted of shells strung in necklaces or on belts. Its use as currency ended when traders began importing exotic tuskshells and plastic or glass beads.

There are 816 fossil and 517 recent species of scaphopods recognized in the world, 41 of which occur in the Gulf of Mexico and 10 in Texas; the most common species is included in this book.

CHAPTER THREE

Common Texas Seashells

Introduction

This chapter describes a total of 300 species of commonly found Texas seashells and illustrates them with more than 600 photographs. These descriptions and photographs are excerpted from our larger book *Encyclopedia of Texas Seashells,* and we refer the reader or user of this field guide who desires more detailed information to see that volume (Tunnell et al. 2010).

Following are brief descriptions of 5 of the 7 seashell classes found in Texas, as well as the more common families. Within each family, we list species alphabetically and include the following information for each: distribution; size, in both metric and English systems; description, in abbreviated or telegraphic writing style; habitat; remarks, or special comments; and, for some, the more commonly recognized synonyms. For a more complete listing of synonyms, please see the *Encyclopedia of Texas Seashells* or visit Gary Rosenberg's Malacolog (www.malacolog.org).

For readers and users of this field guide who desire to learn more about the organization of this book and the collections used for photographs herein, please see Tunnell et al. (2010).

PART I

CLASS: POLYPLACOPHORA
Chitons

The class Polyplacophora, commonly known as chitons, is made up of about 900 species worldwide. Chitons are most commonly found on rocky coastlines, like those of the western United States, where there are many species. Chitons are bilaterally symmetrical marine mollusks that move very slowly. They cling to and graze on rocky shores, usually in shallow coastal waters. They are oblong, flattened mollusks, and their primary characteristic is 8 dorsal, articulating shells or plates that are typically brightly colored and easily visible on most species. To a varying degree these shells are embedded in and surrounded by a fleshy muscular girdle around the edges. The bottom portion of the animal is dominated by a large foot used to move along the rocky habitat. Chitons can be herbivorous or omnivorous and use a well-developed radula (a tonguelike organ with small teeth) to scrape algae or small encrusting animals attached to the rocks. Texas has 5 families of chitons and 7 species. This field guide discusses the 2 families and 2 species most common on the Texas coast.

Ischnochitonidae

Ischnochitonidae is a large and diverse family of chitons made up of approximately 175 species worldwide. Their shape ranges from oval to elongate. The shells are usually convex and with either a low or high elevation. Ornamentation varies from smooth to sculptured. Shell characteristics shared by the individuals of this family are the presence of many slits in the insertion plates of the 2 end shells and only 1 or 2 slits in the intermediate shells. On the Texas coast this family is represented by 2 genera and 3 species. In this field guide we show the single most abundant species of chiton found in Texas.

Ischnochiton papillosus (C. B. Adams, 1845)   **Mesh-pitted Chiton**

Distribution: Florida, Texas; Caribbean.

Size: 10 to 19 mm (⅖ to ⅘ in).

Description: Color greenish. Shape elongate-oval. Ornamentation of numerous small beads arranged in angled rows. Central valves slightly arched; lateral areas weakly raised; girdle covered with tiny ridges of oval scales. **Habitat:** On rocks, shell, or other hard substrate at depths from 0 to 37 m (121 ft). **Remarks:** Found regularly on hard substrates in Texas bays south of Galveston.

Chaetopleuridae

Chaetopleuridae is a small family of chitons made up of only 1 genus and 3 species found in the Gulf of Mexico. The sculpture of the shell surface of this family is radial or randomly arranged with longitudinal beaded riblets on the outer regions of the shell. The 6 intermediate shells between the 2 end shells have only 1 slit on each side where they insert under each other. The girdle is tough, with many spicules. In Texas the family is represented by only 1 species.

Chaetopleura apiculata (Say, 1834)

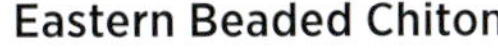

Eastern Beaded Chiton

Distribution: Cape Cod, Massachusetts, Florida, Texas.
Size: 7 to 20 mm (⅓ to ⅘ in).

Description: Color light brownish to yellowish with a lighter or darker coloration in the middle of the valves. Shape broadly oval. Distinct axial carina in center of valves; side areas slightly elevated; head and tail valves with small scattered beads; central valves with angled rows of beads. Girdle thick and leathery with microscopic beads and scattered hair-like spines in fresh specimens. **Habitat:** On other shells or rocks below low-tide line to depths of 37 m (121 ft). **Remarks:** In Texas has been recorded on beach drift in San Luis Pass; live specimens were taken at South Padre Island. Other specimens have been recorded from Heald Bank.

PART II

CLASS: GASTROPODA
Snails, Conchs, and Whelks

At 70,000 living species worldwide there are more gastropods than any other class of mollusks. They are very diverse and numerous. Gastropods range from the primitive deep-water marine species to the complex air-breathing terrestrial snails and slugs. They are found in almost all types of terrestrial, freshwater, and marine environments. This field guide covers only those gastropods associated in some way with the marine environments of Texas. Gastropods may burrow, crawl, swim, float, or be sedentary. They are herbivores, carnivores, omnivores, or parasites. The name Gastropoda, meaning "stomach-foot," comes from Greek (*gastro,* "stomach," and *poda,* "foot"). Gastropods get their name from the fact that their large visceral mass is located on top of their muscular foot. In addition, they generally have a head with eyes, a terminal mouth, and tentacles. Gastropods are unique in that most are asymmetrical as adults but exhibit bilateral symmetry as larvae. The visceral mass and head-foot are often contained in a spiral shell, although many gastropods do not have a shell. The most familiar shell is a hollow cone coiled into a spire, but you will also see cap-shaped species and tubular species in the pages that follow. Texas has 105 families and 593 species of gastropods; this field guide discusses and illustrates 60 families and 156 species that are most common in shallow water.

Fissurellidae

Fissurellidae is a large family of gastropods of about 20 genera usually found in tropical or temperate waters. They are commonly called keyhole limpets, looking something like a small Chinese hat with a hole in the top. Shell shape is from almost flat to conical. Most have a hole at the top of the shell or a slit in the center or front end of the shell. Shell sculpture is variable and can be smooth or highly ornamented with radiating ribs and spiral cords. Some distinct features of the anatomy of these animals are paired hearts, kidneys, and gonads. Most fissurellids are restricted to firm substrates in intertidal and subtidal zones of marine environments with high energy or wave action. The keyhole limpets are not as noticeable as the true limpets, for they tend to shelter themselves underneath rocks and in crevices. Fissurellids are typically herbivores, but some are carnivores feeding on sponges, and others are detritivores. In Texas Fissurellidae is represented by 7 genera and 18 species whose size ranges from 6 to 51 mm (¼ to 2 in); 2 species are discussed here.

Diodora cayenensis (Lamarck, 1822)

Cayenne Keyhole Limpet

Distribution: Maryland to Florida, Texas to Brazil; Bermuda. **Size:** 25 to 51 mm (1 to 2 in).

Description: Color light brown to greenish, usually with dark radiating rays. Shell shape cap-like. Sculpture of radial and spiral ribs, with each fourth radial rib large and 3 smaller radial ribs in between. Orifice or hole on the top of shell keyhole shaped and located in front of and lower than apex. **Habitat:** In Texas found on shelly bottoms and rocks from the intertidal region to about 35 m (115 ft). Occasional to common on Texas jetties. **Remarks:** Shape and height of shell variable. In Texas the maximum size recorded is 27 mm (~1 in). **Synonyms:** *D. alternata* Say, 1822.

Lucapinella limatula (Reeve, 1850)

File Fleshy Limpet

Distribution: North Carolina to Florida, Texas; Caribbean; Brazil.

Size: 6 to 15 mm (¼ to ⅗ in).

Description: Color variable and mottled, from white, yellow, brown, to pink, with some exterior coloration stained through to inner surface. Shell small, flattened, and slightly cap shaped. Sculpture of radiating ribs with ribs originating at orifice or hole on top of shell; spiral sculpture of few larger spiral ribs and numerous smaller threads, creating a beaded appearance where they intersect axial sculpture. Orifice located toward the center of shell. Margin of base thickened and finely crenulate. **Habitat:** On rocks and shells at depths from 0 to 45 m (150 ft). In Texas found alive on oyster shells in bays south of Corpus Christi. **Remarks:** Rare species. In specimens from South Padre Island a blackish ring encircles the interior callus. **Synonym:** *Fissurella aculeata* Dall, 1889.

Calliostomatidae

Calliostomatidae is a widespread family of gastropods found from the intertidal zone down to the deep sea. Their shells range from small to medium and are commonly called top shells. They typically have a beaded spiral sculpture. In Texas Calliostomatidae is represented by 1 genus and 7 species whose size ranges from 8 to 41 mm (⅓ to 1⅗ in); 1 species is discussed here.

Calliostoma euglyptum (A. Adams, 1855)   **Sculptured Topsnail**

Distribution: North Carolina, Florida to Texas; Mexico.
Size: 25 mm (1 in).

Description: Color variable from yellowish to purplish-pink or rose-brown, normally speckled white, and the apex usually dark purplish-brown. Shell shape trochoid and strong, almost as wide as it is high. Sculpture of strongly beaded whorls, made up of 6 beaded spiral cords with 1 weakly beaded spiral thread in between. No umbilicus. **Habitat:** Under hard substrate, from low-tide mark to 55 m (180 ft). **Remarks:** The scientific name comes from the Greek meaning "well sculptured," a good description for its ornamentation. Originally named in the genus _Zizyphinus._ Found living at South Padre Island. **Synonym:** _Zizyphinus euglyptus_ A. Adams, 1855.

Trochidae

Trochidae is a family found worldwide in many kinds of marine environments. Shell shape is variable but is usually spiral/conic or turbinate. Trochids vary in size and shape and can be from low and thin to tall and thick. Sculpture is usually complicated, with varying numbers and strengths of beaded or smooth spiral cords. Ornamentation usually continues onto the base of the shell. The economically valuable larger top shells are the most abundant among tropical reef mollusks, and sturdy temperate trochids are found on rocky shores. Trochids have established themselves as grazers, particularly on brown and red algae. Although many trochids are typically collected while hiding under rocks or in crevices during daylight hours, most are nocturnal herbivores. Trochids use their radula to feed on algae and vegetative detritus. Sexes are separate, and in some species the female is larger than the male. Fertilized eggs may be laid singly or in gelatinous strings or masses. On the Texas coast Trochidae is represented by 2 genera and 7 species whose size ranges from 2 to 25 mm (¹⁄₁₂ to 1 in); 1 species is discussed here.

Tegula fasciata (Born, 1778)   Silky Tegula

Distribution: South Florida, Texas; Caribbean to Brazil.
Size: 12 to 18 mm (½ to ¾ in).

Description: Color reddish-brown with patches of different shades of red, brown, black, and/or white. Shell shape turbinate. Whorls finely sculptured, possessing strong cords with about 3 spiral striae between cords. Umbilicus white. **Habitat:** Common on turtle grass, on algae, and under rocks from low-tide line to a depth of 90 m (295 ft). **Remarks:** Dead shells commonly found in beach drift around Port Isabel and Port Aransas. However, one incidence of a live specimen was reported by P. McGee from Port Aransas. **Synonym:** *T. picta* Tenison Woods, 1877.

Phasianellidae

Phasianellidae is a small family of gastropods made up of a few species found in warm tropical waters among algae, seagrasses, sand, rocks, or coral rubble. A typical phasianellid shell is smooth and ovate to elongate-ovate. The shell appears fragile but is sturdy and has an oval aperture and a glossy surface. Smaller males are found on larger females during mating. In Texas Phasianellidae is represented by 1 genus and 2 species whose size ranges from 5 to 8 mm (⅕ to ⅓ in); 1 species is discussed here.

Eulithidium affine (C. B. Adams, 1850) **Stained Pheasant**

Distribution: Texas to lower Caribbean; Brazil.

Size: 5 to 8 mm (⅕ to ⅓ in).

Description: Color variegated, rose-brownish to rose-greenish base with spirally arranged, somewhat regularly spaced red spots, usually paired with white dots and with dark green, reddish-brown, or black zigzag-type splotches. Shell shape high, conical, and broader at base. Sculpture smoothish. Spire stout. Umbilicus slit-like. **Habitat:** In seagrass beds, primarily turtle grass; also on algae and coral reefs in depths up to 10 m (30 ft). **Remarks:** Found alive at South Padre Island. Shells found dredged as far north as Galveston, offshore at depth of 29 m (96 ft). **Synonym:** _Tricolia affinis cruenta_ Robertson, 1958.

Neritidae

Neritidae is a large family of gastropods comprising approximately 14 genera and hundreds of species. Members of the family Neritidae are found in fresh, brackish, marine, tropical, and subtropical waters worldwide. Few are found in warm temperate climates. The shell is heavy with a low spire. Sculpture may be heavy or smooth. Ornamentation and coloration are usually distinct in many species within this family. The aperture is half-moon shaped. Nerites typically thrive intertidally to just below the low-tide line. Nerites are primarily herbivores, feeding on algae by scraping it off rocks with their wide, rasping radula. In Texas Neritidae is represented by 3 genera and 5 species whose size ranges from 6 to 32 mm (¼ to 1½ in); 4 species are discussed here.

Nerita fulgurans Gmelin, 1791   **Antillean Nerite**

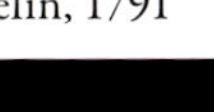

Distribution: Southeast Florida, Texas; Caribbean; Bermuda.
Size: 18 to 32 mm (¾ to 1½ in).

Description: Color varies from black to white with color patterns appearing blurred. Shell shape conic and globose. Body whorl large. Flat columella bearing teeth or folds. Aperture yellowish-gray, circular with 2 distinct teeth located on the inner portion of the outer lip. **Habitat:** Hard substrates like jetties and pilings. Extreme South Texas only. **Remarks:** A peglike projection found in the operculum is inserted into the muscle of the snail for a tight fit when the animal is exposed; similar to *N. tessellata* Gmelin, 1791. However, the spiral ridges are more numerous on *N. fulgurans,* and the aperture is generally wider.

Neritina usnea (Röding, 1798)   **Olive Nerite**

Distribution: Florida to Texas; Caribbean.
Size: 12 mm (½ in).

Description: Color brownish-green to brownish-yellow with numerous striations of black, purple, or brown with a dark almost black band at the suture. Shell shape globose with a large body whorl. Sculpture smooth, glossy, typically with spire eroded away. Aperture half-moon shaped, with parietal area yellowish with numerous variable-sized small teeth.
Habitat: Common in brackish water in muddy, sandy substrates.
Remarks: Entire Texas coast. **Synonyms:** *Nerita usnea* Röding, 1798; *Neritina reclivata* (Say, 1822).

Neritina virginea (Linnaeus, 1758)   **Virgin Nerite**

Distribution: Florida to Texas; Caribbean; Bermuda; Brazil. **Size:** 12 mm (½ in).

Description: Color and mottlings highly variable; dominant color may be olive, white, red, black, or purple with spots, stripes, waves, or lines. Shell small, shape globular. Sculpture glossy smooth with large body whorl. Aperture semilunar with a variable number of small irregular teeth on the inner edge. **Habitat:** Widespread in brackish-water intertidal grassflats. **Remarks:** Known to live in Port Isabel and Port Aransas in inlet-influenced areas. On aerial prop roots of black mangrove (_Avicennia germinans_).

Smaragdia viridis (Linnaeus, 1758)   **Emerald Nerite**

Distribution: Southeast Florida, Texas; Caribbean; Bermuda. **Size:** 6 to 8 mm (¼ to ⅓ in).

Description: Color bright emerald green with white mottlings typically outlined in purplish-black. Shell subglobular, solid. Sculpture glossy smooth. Aperture half-moon shaped with 7 to 9 teeth on inner edge; outer edge thin and sharp. **Habitat:** Typically in seagrass; intertidal at depths to 18 m (60 ft). **Remarks:** Collected in Texas from Port Aransas and southward. True _S. viridis_ comes from the Mediterranean, and some workers separate this form into _S. viridemaris_ Maury, 1917. **Synonyms:** _S. viridis viridemaris_ Maury, 1917.

Cerithiidae

Cerithiidae is the largest family in the superfamily Cerithiodea. Ceriths are found worldwide in all warm and temperate marine waters, particularly in warm, shallow waters in tropical and subtropical climates. Shells of ceriths are high, narrow, and multiwhorled. A usually well-developed siphonal canal is found anteriorly. Ceriths inhabit many substrates, including sandy and/or coral rubble areas of coral reefs. They are also found on rocks and sand flats in subtropical areas. Most species are algivores or detritivores. In Texas Cerithiidae is represented by 2 genera and 5 species whose size ranges from about 5 to 42 mm (⅕ to 1⅔ in); all 5 species are discussed here.

Bittiolum varium (Pfeiffer, 1840) **Grass Cerith**

Distribution: Maryland to Florida, Texas; Brazil.

Size: 5 to 6 mm (⅕ to ¼ in).

Description: Color variable, typically with whitish-tan base stained with darker shades of brown. Shell turret shaped (with a high conic spire). Ornamentation of 7 to 8 convex whorls; axial ribs crossed by spiral ridges, producing small squares and giving a nodulose appearance. Suture incised and distinct. Aperture oval and thin with an adjacent varix. Siphonal canal poorly developed. **Habitat:** On seagrass in bays just below low-tide line to a depth of 11 m (36 ft). **Remarks:** Food for sciaenid or drum-type fishes. **Synonym:** *Diastoma varium* (Pfeiffer, 1840).

Cerithium atratum (Born, 1778)   **Dark Cerith**

Distribution: North Carolina, Florida, Texas; Brazil.
Size: 25 to 38 mm (1 to 1½ in).

Description: Color of base whitish, with spiral brownish bands. Shell shape elongate, turret shaped (with a high conic spire). Sculpture of numerous weak, flat-sided convex whorls, possessing numerous rows of spiral beaded cords on each whorl, with fine granulated threads between the beaded cords. Spire pointed. Aperture oblique with outer lip crenulated into a thick varix. Siphonal canal and anal canal well developed. **Habitat:** Found on limestone, sand, and rubble and on turtle grass in shallow water. **Remarks:** Differs from _C. litteratum_ by its narrower form and smaller beads. Also commonly called the Florida cerith. **Synonym:** _C. floridanum_ Mörch, 1876.

Cerithium eburneum Bruguière, 1792   **Ivory Cerith**

Distribution: Florida, Texas; Yucatán; Brazil; Bahamas.
Size: 43 mm (1⁷⁄₁₀ in).

Description: Color white to yellowish-white with irregular brownish maculations on surface of shell. Shape elongate-conic. Sculpture of shell with numerous spiral nodulose cords; interrupted spiral threads on surface of shell crossed by axial lines, giving a beaded appearance. Spire pointed. Aperture oval. Anal and siphonal canals narrow. Siphonal canal arched backward. **Habitat:** Sandy areas among seagrass beds at depths from 0 to 18 m (60 ft). **Remarks:** Common in shallow water just below the tide line. Adult shells usually lack a protoconch. **Synonym:** _C. algicola_ C. B. Adams, 1845.

Cerithium litteratum (Born, 1778)   **Stocky Cerith**

Distribution: Southeast Florida, Texas; Bermuda; Caribbean; Brazil.

Size: 25 mm (1 in).

Description: Color white with numerous rows of black or reddish-brown squares. Shell broadly elongate-conic. Sculpture of numerous coarse, knobby, spiral cords on each flat-sided whorl. Aperture oblique and oval. Apertural side of last whorl flattened with distinct anal and siphonal canals. **Habitat:** On hard substrates covered with algae, coral rubble, and marine grasses. In Texas found at Flower Garden Banks and Stetson Bank. Found dead at South Padre Island. Depth range 0 to 88 m (289 ft). **Remarks:** *C. litteratum* feed primarily on detritus, algae, and deposits of calcium carbonate.

Cerithium lutosum Menke, 1828   **Variable Cerith**

Distribution: Florida, Texas to Quintana Roo; Costa Rica, Caribbean; Bermuda.

Size: 6 to 21.7 mm (¼ to ⁹⁄₁₀ in).

Description: Color brown, black to grayish-white with brown to reddish mottling. Shell shape elongate-conic. Sculpture of spiral threads between varying strengths of beaded spiral cords; numerous convex, flat-sided whorls. Aperture oval; outer lip white and thickened. Siphonal canal well developed and turns upward. Slight to obscure anal canal. **Habitat:** A littoral species found in seagrasses in shallow bays. **Remarks:** Size frequency data indicate that variable ceriths reach adulthood in approximately one year. **Synonyms:** *C. variabile* C. B. Adams, 1845; *C. versicolor* C. B. Adams, 1850.

Litiopidae

Litiopidae is a distinct family of small cerithioideans that occur in large numbers on algal mats and seagrasses in warm temperate and tropical seas worldwide. Litiopids differ from other cerithioideans by having smooth or poorly sculptured shells. The litiopids have a weakly developed anterior siphonal notch. The sculpture is normally spiral and weak. In Texas Litiopidae is represented by 2 genera and 2 species whose size ranges from 3 to 9 mm (⅛ to ⅓ in); both species are discussed here.

Alaba incerta (d'Orbigny, 1841)   Varicose Snail

Distribution: Florida, Texas; Caribbean; Bermuda; Brazil.
Size: 5 to 9 mm (⅕ to ⅓ in).

Description: Color in fresh shells mottled with axially arranged flammules and brownish spots on a white somewhat translucent background. Shell slender and conic. Sculpture smoothish; numerous somewhat convex whorls, defined by a deep suture, with varices present on several whorls. Spire pointed. Protoconch with axial riblets. Aperture quadrangular with siphonal canal weak, almost not seen. **Habitat:** Primarily found among hard substrates and grass beds. Depth range 0 to 40 m (130 ft). **Remarks:** Characterized by its coloration, acute spire, and varices on last 2 whorls.

Litiopa melanostoma Rang, 1829 **Sargassum Snail**

Distribution: Same as *Sargassum:* North Carolina to Uruguay; Texas.

Size: 3 to 6 mm (⅛ to ¼ in).

Description: Color variable; tan, yellow, and/or brown. Shell shape broadly conic. Sculpture of shell smoothish. Protoconch with axial and spiral sculpture. Body whorl extended, approximately half the size of the shell. Aperture semilunar with convex columella with a ridge at base of columella. Siphonal canal weak. **Habitat:** Pelagic with *Sargassum* but is commonly found washed ashore in beach drift. Depth range 0 to 805 m (2640 ft). **Remarks:** Differs from *Alaba incerta* by being shorter and wider with fewer whorls and a taller aperture.

Potamididae

Potamididae is a moderate-sized family of gastropods of about 12 genera and many species. Potamidids inhabit brackish waters in tropical and subtropical regions of the world. Potamidids are mud dwellers, whereas ceriths reside in sandy substrate. The aperture is oval to angled with a degree of flaring on the outer portion of the lip and a clearly marked siphonal canal. On the Texas coast Potamididae is represented by 1 species.

Cerithidea pliculosa (Menke, 1829)   **Plicate Hornsnail**

Distribution: Texas, Louisiana; Caribbean.

Size: 12 to 33 mm (½ to 1⅓ in).

Description: Color base brownish-black, with varices yellow or orange-yellow with a yellow band at center of each whorl. Shape elongate-conic. Sculpture of 11 to 13 convex whorls. Axial ribs irregular and strong. Spiral sculpture absent. Presence of numerous varices a characteristic feature of mature snails. Aperture subcircular; outer lip thick and forms an outer varix around it. Siphonal canal not well developed. **Habitat:** Found in bay shorelines on mudflats, often near salt marsh cordgrass at depths from 0 to 2 m (7 ft). **Remarks:** Juveniles without varices easily confused as ceriths.

Modulidae

Modulidae is a small family of snails that is represented by 1 genus, *Modulus,* and 6 species. Most species live in shallow, subtidal, warm-water seagrass beds; however, some species may be found underneath coral boulders. The shell of modulids is low spired and robust. Sculpture is spiral and has 5 to 6 whorls. Modulids are basically herbivores that feed on small plant life and detritus. In Texas Modulidae is represented by 1 species.

Modulus modulus (Linnaeus, 1758) · **Buttonsnail**

Distribution: North Carolina to Texas; Brazil; Bermuda.
Size: 10 to 16 mm (⅖ to ⅗ in).

Description: Color yellowish-white with brownish-red axially arranged splotches or bands on spiral cords. Shell shape conic and robust. Shell sculpture of heavily beaded spiral cords. Spire moderately low. Body whorl large with keeled base. Upper part of shell marked with spiral ridges and transverse growth lines; shell has 5 strong cords at base; deeply incised suture at periphery. Umbilicus slit-like and deep. Aperture round with a thick crenulate outer lip with a distinct tooth on the inner side of the apertural opening. **Habitat:** In Texas typically found in beach drift. Depth range 0 to 105 m (344 ft). **Remarks:** Slipper shells are often attached to these shells.

Scaliolidae

Scaliolidae (= Finellidae) is a small family of mollusks found through-out the tropical and subtropical shallow-water regions of the world. The size of shells ranges from 2 to 4 mm (⅓ to ⅙ in). Shells are towerlike and elongate. Sculpture may be spiral and/or axial, which is cancellate in _Finella,_ or without sculpture and cemented with sand grains. In Texas Scaliolidae is represented by 1 genus and 2 species whose size is approximately 3 mm (⅛ in); 1 species is discussed here.

Finella dubia (d'Orbigny, 1840)

Dubious Finella

Distribution: North Carolina to Florida, Texas; Brazil; Bermuda.

Size: 3 mm (⅛ in).

Description: Color translucent white. Shape elongate-conic. Sculpture of axial and spiral ribs giving cancellate appearance. Protoconch smooth. Aperture rounded with a thin outer lip and weakly curved columella. **Habitat:** Found in intertidal and shallow, subtidal sandy-mud-type substrates to depths up to 805 m (2640 ft). **Remarks:** This snail is difficult to distinguish from *Bittiolum varium* and *Cerithium lutosum*. **Synonyms:** *Alabina cerithidioides* (Dall, 1889); *Bittium cerithidioides* (Dall, 1889).

Turritellidae

Turritellidae is a large family of gastropods found in marine environments worldwide. These gastropods live from the subtidal zone to deep water, on soft muddy and/or sandy bottoms, in gravel, in shell grit, and sometimes wedged among rocks and crevices. In some locations they are the most abundant element that makes up the bottom. Shells are small to large and with no, or a weak, siphonal notch. Turritellids are suspension feeders but may supplement their diet by deposit feeding. Turritellids have a multiwhorled, towerlike shell. On the Texas coast Turritellidae is represented by 2 genera and 4 species whose size ranges from 51 to 90 mm (2 to 3½ in); 1 species is discussed here.

Vermicularia spirata (Philippi, 1836) — West Indian Wormsnail

Distribution: Florida, Texas; Caribbean; Bermuda.

Size: Apex about 6 mm (¼ in); apex and wormlike teleoconch about 90 mm (3 in).

Description: Color of apex grayish; rest of shell brownish-gray. Shell shape irregular, apex conic to pointed; rest of shell wormlike. Whorls detached and appear wormlike. Aperture subcircular and thin lipped. **Habitat:** In Texas found partially embedded in sponges and attached to the tree coral *Oculina.* Depth range 3 to 80 m (10 to 262 ft). **Remarks:** Similar to *V. knorrii;* however, spiral cords are more numerous and smaller on *V. spirata.*

Littorinidae

Littorinidae is a large family of gastropods made up of approximately 100 species found worldwide in cold to tropical waters. Commonly known as winkles or periwinkles, these snails are usually found on hard substrate in the littoral fringe zone. Shells are small to medium sized and turbinate to conic in shape. The shells are from 2 to 50 mm (¹⁄₁₂ to 2 in) in length, solid, and turbinate to conic in shape. Color is variable, often consisting of axial rays or bands. Species found on mangrove foliage are usually yellow, brown, and red. Littorinids feed on microscopic algae film, epiphytes, and leaf hairs from mangrove trees. In Texas Littorinidae is represented by 2 genera and 5 species whose size ranges from 6 to 25 mm (¼ to 1 in); 4 species are discussed here.

Echinolittorina placida Reid, 2009   **Interrupted Periwinkle**

Distribution: Texas to Florida; Caribbean.

Size: 9 to 25 mm (⅓ to 1 in).

Description: Color dark brown at spire, gradually becoming whitish with dark brown to black zigzag markings at angles to the suture. Shell shape elongate-conic. Sculpture of irregular, tightly spaced incised spiral ribs, and usually with 6 to 8 convex whorls. Body whorl more than half the height of shell. Aperture oval to subcircular; outer lip edge thin and sharp. **Habitat:** Found near high-tide line on hard substrates; common in colonies in rock crevices of Texas jetties. **Remarks:** Larger on sandstone jetties than on granite jetties. Taxonomically, hard-shore striped littorines have been problematic. For more information on the taxonomy of this striped littorine, please see Tunnell et al. (2010: 138). Specimen 1 was collected on the bay sandstone jetties of McGee Beach, Corpus Christi (December 2008), and was the largest of 7 specimens collected. It is approximately 16 mm (⅗ in) in length. Specimen 2 was collected on the granite-type jetties of Port Aransas (December 2008) and was the largest of 35 specimens collected. It is about 5 mm (⅕ in) in length. Although different in size and shell characteristics, these specimens are considered to be the same, but variable. At present all Gulf of Mexico striped littorines are considered to be the species _E. placida_ Reid, 2009. However, new research using DNA evidence suggests this "species" to be a species complex, and name changes may be forthcoming.

Littoraria angulifera (Lamarck, 1822)

Mangrove Periwinkle

Distribution: Florida to Texas; Brazil; Bermuda.

Size: 25 mm (1 in).

Description: Color variable, yellowish-white, orange-yellow, or grayish-brown with dark brown axially arranged splotches. Shape conic and broad. Sculpture of numerous spiral threads and somewhat irregular diagonal growth ridges. Shell thin but strong. Body whorl may be carinate and about half total length of shell. Aperture broad and oval. **Habitat:** In calm and brackish mangrove areas above high-tide mark. **Remarks:** In Texas found on hard substrates and bridge abutments. **Synonyms:** *Littorina lineata* Gmelin, 1791; *L. angulifera* (Lamarck, 1822).

Littoraria irrorata (Say, 1822)

Marsh Periwinkle

Distribution: New Jersey to Florida to Texas.

Size: 25 mm (1 in).

Description: Color whitish-gray with streaks of purple or reddish-brown on spiral cords. Shell shape elongate-conic; thick. Sculpture with numerous rounded, somewhat thickened spiral ribs, with 8 to 10 flat-sided whorls; sutures somewhat obscure; body whorl about half the height of shell. Aperture oval, with outer lip strong and sharp. **Habitat:** Typically on marsh grass stems and roots at or above high-tide mark; also on hard substrates at or above tide line. **Remarks:** Sometimes confused with *L. nebulosa,* which has a convex body whorl, different columella, and finer spiral ribs. **Synonym:** *Littorina irrorata* (Say, 1822).

Littoraria nebulosa (Lamarck, 1822) **Cloudy Periwinkle**

Distribution: Texas to Florida; Caribbean to Brazil; Bermuda.
Size: 6 to 28 mm (¼ to 1 ⅛ in).

Description: Color whitish-yellow with light almost obsolete reddish-brown maculations forming zigzag bands on body whorl; spire darker brown on older specimens (specimen 1; size: 25 mm; 1 in); reddish-brown splotches on grayish background for smaller specimens (specimen 2; size: 6 mm; ¼ in). Shell shape conic and broad. Sculpture of rounded whorls and spiral ribs with lower ribs of body whorl becoming bifurcate in older specimens (specimen 1). Smaller specimens with narrow, incised spiral ribs (specimen 2). Body whorl rounded, ⅔ height of shell. Protoconch whorls translucent brown, smooth, and bulbous when intact. Suture distinct. Aperture broad and oval to teardrop shaped; outer lip edge slightly crenulate, thin, and sharp and meets body whorl at angle; columella with a platelike structure that terminates at base of aperture. **Habitat:** Found on jetties and other hard substrates in higher-salinity waters. **Remarks:** Specimen 1 adult *L. nebulosa*. Prefers wooden structures; usually avoids high-energy zones. Specimen 2 from jetties at Packery Channel, Corpus Christi. Young specimens of *L. nebulosa* differ in appearance from mature specimens and have been misidentified as *L. tessellata* (Philippi, 1847) (D. Reid, pers. comm., 2009). **Synonyms:** *Phasianella nebulosa* (Lamarck, 1822); *Littorina nebulosa* (Lamarck, 1822).

Assimineidae

Assimineidae is a fairly large family of gastropods of approximately 12 genera and hundreds of species. Most of these species are located in Asia and the Pacific Islands. Assimineids are amphibious snails associated with marine, freshwater, and land environments. Shells of assimineids are small, ovate to conic in shape, often with spiral color bands. The spire may be high or low. The assimineids are probably detritus feeders, feeding on the organic material found in the sediments where they live. In Texas Assimineidae is represented by 1 species.

Assiminea succinea (Pfeiffer, 1840) **Atlantic Assiminea**

Distribution: Massachusetts to Texas; Quintana Roo; Brazil. **Size:** 2 to 5 mm ($\frac{1}{12}$ to $\frac{1}{5}$ in).

Description: Color pale translucent brown. Shape broadly conic. Sculpture smoothish with 5 to 5¾ rounded whorls. Body whorl makes up about ¾ of shell. Aperture teardrop shaped. **Habitat:** In moist, sun-protected areas of salt marshes and underneath rocks on jetties. Usually found at depths from the surface to about 4 m (13 ft). **Remarks:** Lives just at the water's edge. Fresh specimens have a straw-colored or yellowish coloration. Tiny shell not easily visible to the naked eye.

Caecidae

Caecidae is a large family of gastropods found primarily in tropical waters worldwide. Shells are minute, curved, tubular, and tooth or sausage shaped; they are sometimes called tuskshells. Shell sculpture is variable and may be smooth and glossy, with spiral or axial sculpture, or with both spiral and axial sculpture. Caecids feed on small particles and unicellular organisms that live on sand grains or small pebbles. On the Texas coast Caecidae is represented by 2 genera and 16 species whose size ranges from 1 to 12 mm ($\frac{1}{25}$ to $\frac{1}{2}$ in); 4 species are discussed here.

Caecum imbricatum Carpenter, 1858   Imbricate Caecum

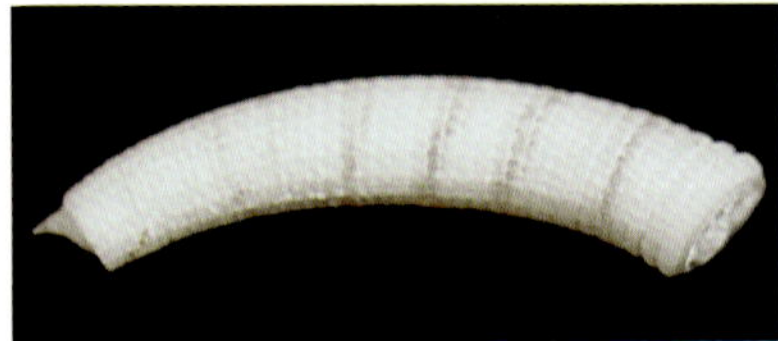

Distribution: Florida to Texas; Bahamas; Caribbean.
Size: 4 to 5 mm (⅙ to ⅕ in).

Description: Color variable, may be whitish, tannish, or a mottled appearance of white and tan with brown bands. Shell shape toothlike or tusklike and strongly arched. Sculpture of both spiral and longitudinal ridges, giving a distinct cancellate appearance. Aperture made up of larger ridges, producing a varix. Mucro (pointed, posterior part of shell) distinctly pointed and not recessed. **Habitat:** In Texas found on sand or shell bottoms at depths from 0 to 83 m (272 ft). **Remarks:** Probably most closely related to *C. floridanum.* **Synonyms:** *C. coronatum* Folin, 1867; *C. formulosum* Folin, 1868.

Caecum johnsoni Winkley, 1908   Johnson's Caecum

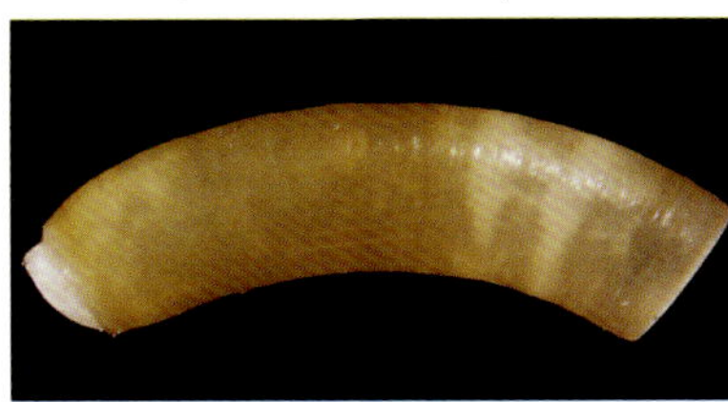

Distribution: Massachusetts, North Carolina, Texas.
Size: 5 mm (⅕ in).

Description: Color may be translucent white or brown. Shell shape curved and tubular, toothlike or tusklike. Sculpture smoothish. Mucro present as knob on side of hemisphere. Aperture smooth and without a varix. **Habitat:** In beach drift along Texas coast. Also in bays, oyster reefs, and offshore reefs and banks from shoreline to a depth of 75 m (246 ft). Alive at Flower Garden Banks. **Remarks:** Sometimes mistaken for *C. glabrum* Montagu, 1803, a European species. Offshore specimens may be smaller.

Caecum pulchellum Stimpson, 1851   **Beautiful Caecum**

Distribution: New Hampshire to Brazil; Texas.

Size: 2 mm (1/12 in).

Description: Color light brown to white. Shell shape toothlike or tusklike and slightly arched. Sculpture with crowded to somewhat spaced spiral ridges, which vary in size and distance between annulations. Mucro a low, triangular projection. Aperture constricted and round. **Habitat:** Seagrass beds, bays, and lagoons. In Texas restricted to offshore shell bottoms and coral reefs. Depth range 1 to 101 m (3 to 330 ft). **Remarks:** Similar to C. *textile;* however, C. *textile* is more arched, and annulations are rounder and smoother. **Synonym:** *C. capitanum* Folin, 1874.

Meioceras nitidum (Stimpson, 1851)   **Little Horn Caecum**

Distribution: South Florida, Texas; Gulf of Mexico to Caribbean; Brazil.

Size: 2 to 3 mm (1/12 to 1/8 in).

Description: Color translucent with irregular mottlings of white. Shell shape broadly toothlike or tusklike. Sculpture smooth, narrow posteriorly, wider in the middle, and compresses at the aperture. Middle part of shell shape variable; may be slightly concave to almost straight to slightly convex. Mucro bluntly pointed and triangular. **Habitat:** Dredged offshore in green algae; shallow bays and lagoons where the salinity remains close to that of seawater. Found from shoreline to a depth of 24 m (79 ft). **Remarks:** Resembles *Caecum cornucopiae* Carpenter, 1858, which is not swollen. Moore (1972) states *C. cingulatum* Dall, 1892 may be a form of *C. nitidum* with a highly inflated middle. **Synonym:** *C. nitidum* Stimpson, 1851.

Hydrobiidae

Hydrobiidae is a large family of many genera and numerous small species of gastropods. Most are freshwater snails, but some are found in brackish water. Hydrobiid shells are diverse in shape. Food sources for hydrobiids include unicellular algae, bacteria, and detritus. Hydrobiids inhabit a wide variety of habitats from estuarine mudflats to caves, groundwater, mountain lakes, rivers, and desert artesian springs. On the Texas coast Hydrobiidae is represented by 4 genera and 6 species whose size ranges from 2 to 4 mm (¹⁄₁₂ to ⅙ in); 3 species are discussed here.

Probythinella protera Pilsbry, 1953

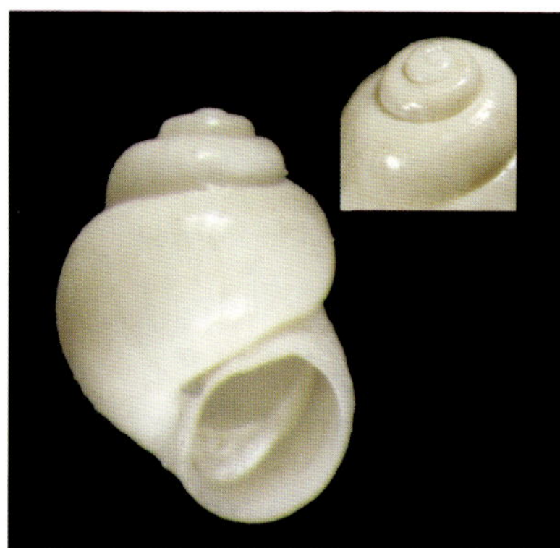

Distribution: Alabama, Mississippi, Louisiana, Texas.
Size: 3 mm (⅛ in).

Description: Color opaque white. Shell shape broadly conic to cocoon shaped. Sculpture smoothish; shell has 4½ convex whorls. Sutures between whorls distinct. Body whorl globose and approximately ¾ height of shell. Aperture oval and separated from the body whorl. Umbilicus chinklike. **Habitat:** Brackish water in upper bays at depths from surface to 2 m (7 ft). **Remarks:** Listed as *Vioscalba louisianae* Morrison, 1965 in Andrews (1977). **Synonym:** *V. protera* Pilsbry, 1953.

Texadina barretti (Morrison, 1965)   **Boone Hydrobe**

Distribution: Mississippi to Texas.

Size: 2 mm (¹⁄₁₂ in).

Description: Color grayish with a darker-colored band beneath the suture. Shell shape elongate-conic. Sculpture smooth with approximately 5½ whorls; whorls distinguished by sutures of unimpressed divisions. Apex subtruncate. Aperture typically teardrop form. **Habitat:** Mud bottoms; in areas influenced by rivers at depths from 0 to 2 m (7 ft). **Remarks:** When the animal is alive, the most conspicuous character is the pinkish coloration of the body. Originally placed in the genus *Odostomia*. **Synonyms:** *Hydrobia barretti* (Morrison, 1965); *O. barretti* Morrison, 1965; *H. boonea* (Morrison, 1968).

Texadina sphinctostoma (Abbott and Ladd, 1951) **Narrowmouth Hydrobe**

Distribution: Mississippi to Texas to Quintana Roo.

Size: 3 mm (⅛ in).

Description: Color translucent grayish-white. Shell shape elongate-conic to fusiform. Sculpture smoothish with 5 to 7 convex whorls. Suture moderately impressed with a darker band just below the suture. Body whorl globose at center and tapers anteriorly. Aperture oval to round, askew, and constricted. **Habitat:** Soft mud in shallow, low-salinity bays at depths from 0 to 4 m (13 ft). **Remarks:** Lives in association with *Rangia cuneata* (Gray, 1831). Most commonly seen hydrobiid on the Texas coast.

Rissoidae

Rissoidae is a large family of gastropods with many species whose classification is still under study. Members of the Rissoidae have a worldwide distribution, and the many known species are quite small. Shells are conic, broadly ovate to narrowly elongate, and smooth to variously sculptured. Rissoids are found in sand, in fine gravel, under rocks, on algae, or on marine plants, primarily in crevices or sheltered areas on rocky shores. These gastropods feed primarily on diatoms and pieces of algae. Because of their small size, only a very few life histories of this family are known. On the Texas coast Rissoidae is represented by 7 genera and 14 species whose size ranges from 1 to 11 mm (¹⁄₂₅ to ²⁄₅ in); 3 species are discussed here.

Rissoina cancellata Philippi, 1847   **Cancellate Risso**

Distribution: Florida, Texas; Caribbean; Brazil; Bermuda.

Size: 5 to 7 mm (¹⁄₅ to 3/10 in).

Description: Color white, with a light yellowish-tan band sometimes seen below the suture in fresh shells. Shell shape elongate-conic. Sculpture formed by distinct elevated spiral and axial ribs, becoming knobby where they intersect, giving distinct cancellate appearance. Suture deeply excavated. Shell with small, smooth nucleus, typically made up of 2½ whorls. **Habitat:** Sand and shelly bottoms at depths from 1 to 143 m (3 to 469 ft). **Remarks:** Common, widespread species found along the Texas coast in many offshore environments.

Schwartziella catesbyana (d'Orbigny, 1842)   Catesby's Risso

Distribution: North Carolina, Florida, Texas to Brazil.
Size: 3 to 4 mm (⅛ to ⅙ in).

Description: Color white. Shell shape elongate-conic. Sculpture usually of slightly rounded whorls that have 11 to 14 smooth and slender axial cords. Protoconch small, pointed, set away from teleoconch. Aperture oval, channeled posteriorly, round anteriorly; parietal area flared back. Adults with prominent tooth on inside of highly thickened outer lip. **Habitat:** South Texas bays at depths from 0 to 40 m (130 ft). **Remarks:** Able to penetrate into brackish water. A herbivore that feeds on benthic microflora. Often confused with *S. chesnelii* (Michaud, 1830); however, *S. chesnelii* has a larger protoconch, no tooth on outer lip, and anatomical differences. **Synonym:** *Rissoina catesbyana* d'Orbigny, 1842.

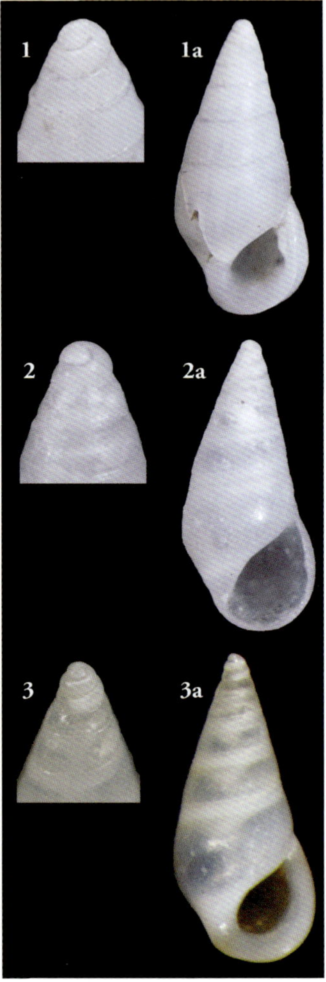

Distribution: Bermuda; North Carolina to Gulf of Mexico, Texas; Caribbean.

Size: 4 mm (⅙ in).

Description: Color translucent white to opaque. Shell shape elongate-conic. Sculpture smooth, flat sided, with approximately 7 slightly convex glassy whorls. Suture shallow, seen just below the surface of the shell. Apex acute. Aperture oval and outer lip usually thickened. **Habitat:** Under shell and debris in grassy bottoms beyond littoral zone. Depth range 0 to 51 m (168 ft). **Remarks:** Confusion exists between offshore and nearshore species or forms. Odé (1986) describes those *Z. browniana* collected offshore in calcareous environments with a tooth located on the inner side of the outer aperture as probably *Z. eulimoides* (d'Orbigny, 1842); whereas those without a tooth found near shore are *Z. browniana*. However, specimens found at the Flower Garden Banks are both toothed and untoothed.

Tornidae

Until recently, this family was known as Vitrinellidae. For more information on the taxonomy of the Tornidae, see Tunnell et al. (2010). The Tornidae is a heterogeneous group of minute marine to brackish-water snails, and its members are among the smallest gastropods known. Shell color in tornids is typically off white and translucent in living and fresh specimens and opaque in dead specimens. Shell shape is generally round and compressed with a somewhat flattened spire. The shell is usually smooth but may have spiral and/or axial sculpture. The protoconch may be multispiral, smooth, or almost smooth to completely covered, as in some teinostomes. Tornids usually possess a simple, rounded aperture and a round umbilicus; however, in some cases, such as species in the genus *Teinostoma,* the umbilicus is usually partially or entirely covered over by a callus. Primary food type for tornids is believed to be detritus and microscopic plant life. In Texas Tornidae is represented by 11 genera and 32 species whose size ranges from 1 to 10 mm (⅕ to ⅖ in); 7 species are discussed here.

Anticlimax pilsbryi (McGinty, 1945)

Cupola Vitrinella

Distribution: Florida to Texas.
Size: Diameter 2.6 mm (⅛ in); altitude 1.7 mm (1/12 in).

Description: Color whitish in fresh material. Shell dome shaped and wider than high. Sculpture of spiral zigzag lines; keel found at periphery; base flat with spiral threads and thickened radial waves. Spire abrupt with protoconch of 1½ whorls and teleoconch of 2½ whorls. Umbilicus usually subcircular and typically bound by a thick ridge and a heavy callus that originates from the columella; callus sometimes reduces the umbilicus to a slitlike structure. Aperture oblique with a thickened outer lip. **Habitat:** Shallow- to deep-water species living on sand or muddy bottoms. Depth range 0 to 152 m (500 ft). **Remarks:** Variation in size and extent of the umbilical callus. **Synonym:** *Teinostoma pilsbryi* McGinty, 1945.

Circulus texanus (Moore, 1965)   **Texas Vitrinella**

Distribution: Texas.
Size: Diameter 1.72 mm ($\frac{1}{12}$ in); altitude 0.78 mm ($\frac{1}{45}$ in).

Description: Color translucent white in live or fresh material to opaque in dead specimens. Shell shape discoid and thin, not planispiral. Sculpture of many spiral grooves with wavy axial growth lines and low, radiating, thickened, wavy riblets on flattened base. Protoconch made up of 1¾ smooth whorls with a varix at the termination of the protoconch. Spire flat and depressed. Umbilicus wide and somewhat deep. Aperture oblique to subquadrate. **Habitat:** Shallow coastal waters at depths from 0 to 44 m (144 ft); may be endemic to the northwestern Gulf of Mexico. **Remarks:** Appears to have a stenotypic (narrow ranges of tolerance type) lifestyle. **Synonym:** *Vitridomus texanus* (Moore, 1965); *Vitrinella texana* Moore, 1965.

Cochliolepis parasitica Stimpson, 1858   **Parasitic Scalesnail**

Distribution: South Carolina to Texas; Caribbean.
Size: Diameter 3 mm ($\frac{1}{8}$ in); altitude 1 mm ($\frac{1}{25}$ in).

Description: Color translucent white in live or fresh material to opaque in dead specimens. Shell shape discoid, not planispiral; shell thin and compressed. Sculpture smooth with the exception of irregular, wavy axial and sometimes spiral growth lines. Spire flat or sunken and partially covered with a thin, translucent callus; rest of shell with 2 adpressed whorls. Umbilicus wide and shallow. Aperture oblique to ovate. **Habitat:** Found beneath the scales of the annelid worm (*Acoetes lupina*). Depth range 0 to 48 m (157 ft). **Remarks:** The distinct growth lines give *C. parasitica* a nautiloid appearance, and the color of the living animal is red. The parasitic name is misleading because the animal lives commensally with the tube worm, feeding on algae and/or detritus found within the elytra of the polychaete annelid. **Synonym:** *Adeorbis nautiliformis* Holmes, 1859.

Cochliolepis striata Dall, 1889 **Striate Scalesnail**

Distribution: Florida to Texas. **Size:** Diameter 6.5 mm ($\frac{7}{10}$ in); altitude 2.0 mm ($\frac{1}{12}$ in).

Description: Color translucent white in live or fresh material to opaque in dead specimens. Shell shape discoid, not planispiral; shell thin. Sculpture of many fine spiral threads and numerous wavy, irregular axial growth lines. Spire flat with the first whorl on the protoconch smooth, and remaining whorls of protoconch covered with a thin, translucent callus. Umbilicus constricted with numerous fine radial growth lines. Aperture shape oblique to ovate. **Habitat:** In Texas it is probably widely distributed, as shells have been found in beach drift all along the Texas coast. Depth range 0 to 46 m (150 ft). **Remarks:** *C. striata* is thought to live commensally with a polychaete host, such as *C. parasitica.*

Episcynia inornata (d'Orbigny, 1842) **Fringed Vitrinella**

Distribution: North Carolina, south Florida, west coast of Florida, Texas; Puerto Rico; Virgin Islands; Panama.
Size: Diameter 3.4 mm (⅛ in); altitude 2.0 mm (1/12 in).

Description: Color translucent white in live or fresh material to opaque in worn or dead specimens. Shell shape trochoid; shell thin and fragile. Sculpture of a low, narrow peripheral keel with many projecting variable-sized teeth, with faint wavy axial growth lines on top and base of body whorl. Protoconch slightly more than 1 whorl, but not as depressed as in most tornids. Umbilicus narrow, flat sided, deep, and bordered by a keel. Aperture flattened and oval. **Habitat:** In sand, mud, and rocky outcrops. This tornid is found all along the Texas coast from the shoreline to a depth of about 128 m (420 ft). **Remarks:** Uncommon but widespread shallow-water species. **Synonym:** *Vitrinella multicarinata* Dall, 1889.

Teinostoma biscaynense Pilsbry and McGinty, 1945 **Biscayne Vitrinella**

Distribution: Florida to Texas.
Size: Diameter 1.9 mm (1/12 in); altitude 0.9 mm (1/25 in).

Description: Color translucent white in live or fresh material to opaque in dead specimens. Shell shape trochoid, depressed, and smooth. Sculpture of faint axial growth lines and faint to obscure spiral threads. Spire covered by a thin, glazed callus. Umbilical callus fills the entire umbilical depression in mature specimens. Aperture oblique. **Habitat:** Inlet areas and surf zone. Shells found on Texas beaches from shoreline to a depth of 1 m (3 ft). Depth range 0 to 500 m (1640 ft). **Remarks:** *T. biscaynense* is different from other *Teinostoma* species by having a spire that is glazed over. **Synonym:** *T. nesaeum* Pilsbry and McGinty, 1945.

Vitrinella floridana Pilsbry and McGinty, 1946   Florida Vitrinella

Distribution: Southeast Florida to Texas.
Size: Diameter 2 mm (¹⁄₁₂ in); altitude 1 mm (¹⁄₂₅ in).

Description: Color translucent white in live or fresh material to opaque in worn or dead specimens. Shell shape somewhat discoid to almost flattened. Sculpture smooth except for weak axial growth lines. Protoconch projects slightly with 2¼ whorls. Teleoconch of 2 whorls delineated from protoconch by the suture. Umbilicus smooth, wide, and deep. Aperture oblique to subquadrate. **Habitat:** Inlet-influenced areas and near shore; in all bays of Texas. Depth range 0 to 46 m (150 ft). **Remarks:** This is basically a featureless shell. Resembles *V. helicoidea;* however, the umbilicus of *V. helicoidea* is not wide but deep and carinate at base of shell.

Truncatellidae

Truncatellidae is a relatively small family of tropical gastropods made up of about 6 genera and few species found throughout the world. Truncatellids are semiterrestrial to terrestrial snails with cylindrical, truncate, high-spired shells that may be sculptured or unsculptured. Shells with smooth and/or ribbed sculpture may be found in a single population. It is thought that the truncatellids feed on detritus and small unicellular plants because their alimentary canal is quite similar to that of the rissooideans. Most truncatellids inhabit areas at or just above the tide line in warm temperate and tropical areas. On the Texas coast Truncatellidae is represented by 1 species.

Truncatella caribaeensis Reeve, 1842   **Caribbean Truncatella**

Distribution: Florida to Texas; Caribbean; Bermuda.
Size: 6 to 8 mm (¼ to ⅓ in).

Description: Color white to pale amber. Shell shape elongate-cylindrical. Sculpture with several axial ribs fading at the base, with 4 to 4½ whorls; many poorly developed transverse costae found primarily on top part of whorls. Mature specimens very different from juveniles due to loss of early whorls. Aperture ovate and slightly flaring. **Habitat:** Sandy bottoms along the shore; sometimes in bays and inlets. **Remarks:** Found in Texas from Port Aransas and southward. Erroneously called *T. pulchella* Pfeiffer, 1839.

Strombidae

Strombidae is a moderate-sized family of gastropods. The strombids inhabit tropical and temperate waters worldwide. Shells range in size from medium to large, reaching 400 mm (16 in) in length. These shells have a broad, conic, and pointed spire that may be smooth or sculptured. Strombid shells are heavy and have a moderate to high spire with an enlarged last whorl. The aperture on strombids is long. The outer lip is typically flared outward. Strombids have a thick, expanded apertural outer lip with a distinct notch near the lower end ("stromboid notch"), enabling the right eye to extend outward. Sexes are separate, with males significantly smaller than females. Juveniles are quite different with taller spires and unflared apertures. Strombids are herbivores that feed on macroalgae and epiphytes. These animals are edible, and a huge fishery once thrived for the queen conch (*Eustrombus gigas*) in the Caribbean and southern Gulf of Mexico but has since dwindled due to overfishing. In Texas Strombidae is represented by 4 genera and 5 species whose size ranges from 40 to 305 mm (1/F>⅗ to 12 in); 1 species is discussed here.

Strombus alatus Gmelin, 1791   **Florida Fighting Conch**

Distribution: North Carolina to east and west coasts of Florida to Texas.

Size: 70 to 102 mm (3 to 4 in).

Description: Color from salmon-pink, orange-brown, to brownish-red. Shell shape somewhat broadly conic. Sculpture of 9 to 10 whorls with rough axial ridges on whorls immediately following the protoconch; at about the fourth whorl the ridges pinch up and form a row of spines, which are longest on the last 2 whorls; axial sculpture of fine threads. Spire with 8 whorls. Body whorl about ⅘ height of shell. Aperture long and narrow with a thickened, outward-flaring lip. Parietal wall and inner portion of outer lip polished and vibrantly colored salmon-pink. **Habitat:** Intertidal in marine grasses to a depth of approximately 183 m (600 ft).

Remarks: This species is sometimes found in an albino form with or without spines.

Calyptraeidae

Calyptraeidae, formerly known as Crepidulidae, is a relatively small family of gastropods of about 60 species found in warm and cold waters. Shell outline is limpetlike, caplike, or slipper shaped. Calyptraeids can be coiled and low spired, ear shaped, or patelliform. All show coiling at the apex. Most shells exhibit no sculpture; however, some may. Although these snails are able to move, this movement is best described as a creeping motion. Most species are protandrous hermaphrodites. Calyptraeids live primarily in subtidal zones in muddy bays, in empty shells, or on any other hard substrate. In Texas Calyptraeidae is represented by 3 genera and 6 species whose size ranges from 6 to 50 mm (¼ to 2 in); 3 species are discussed here.

Crepidula convexa Say, 1822   **Convex Slippersnail**

Distribution: Massachusetts to Florida, Texas; Caribbean; Bermuda.
Size: 6 to 12 mm (¼ to ½ in).

Description: External color varies from tan to purple with reddish lines distinct on tan shells, with some shells having brown spots or maculations on a white base; interior of shell chestnut brown to bluish-brown. Shell shape limpetlike. Outer sculpture smooth with fine radial growth lines. Inner portion of shell has a septum (deck) that covers less than half the aperture and is slightly longer on left than on right side of shell. Midline almost straight. Muscle scar located on right side just beneath right corner of the deck. Apex centrally located, near posterior margin. Aperture large and oval with a thin margin. **Habitat:** Intertidally on shell, rocks, and grass to moderate depths. Depth range 0 to 70 m (230 ft). **Remarks:** These snails are regularly found living on _Argopecten irradians amplicostatus_ (Dall, 1898).

Crepidula depressa Say, 1822   **Eastern White Slippersnail**

Distribution: Canada to Texas; Brazil; Bermuda.
Size: 12 to 37 mm (½ to 1½ in).

Description: Color uniformly white or tinged yellow. Shell shape flat, elongate-oval. Sculpture smoothish. Septum (deck) is convex, shallow, and covers approximately half of aperture. Apex located centrally at posterior end is pointed, directed posteriorly, and detached from rest of shell. Aperture large, oval, and thin. **Habitat:** Intertidal at depths from 0 to 3 m (10 ft). Usually found attached to the inside walls of empty _Polinices_ (moonsnail) shells.

Remarks: *C. depressa* has been misidentified as *C. plana* Say, 1822. Collin (2002) restricts *C. plana* to Georgia and northward. *C. depressa* has shell characters identical to those of *C. unguiformis* Lamarck, 1822; however, *C. depressa* is a western Atlantic species, and *C. unguiformis* is a Pacific species.

Crepidula fornicata Linnaeus, 1758 — **Common Atlantic Slippersnail**

Distribution: Canada to Florida, Texas.

Size: 18 to 50 mm (¾ to 2 in).

Description: Color of alternating white and brown irregularly shaped rays, usually with a broad white ray toward the middle; interior tan to purple; septum (deck) with a brown ring around a white deck. Shell shape varies, round to ovate. Sculpture smoothish. Body whorl primary component of shell. Deck covers more than half of aperture with left side longer than right. Concave, shallow, muscle scar is absent. Apex blunt, twisted right, adherent to margin of shell. Aperture oblong, oval with a thin margin. **Habitat:** Typically a littoral species with individuals growing on one another. Depth range 0 to 70 m (230 ft). **Remarks:** Because of the variability of this species, it is surprising there are not more synonyms.

Cypraeidae

Cypraeidae is a large family of gastropods, also known as cowries, with about 220 species found in tropical waters of the world, but most diverse in the Indo-Pacific. The shell is smooth and brightly colored with dentition of adult shells found on the inner and outer lips of the apertural opening. Current classification is based on shell shape and apertural dentition, but molecular studies are shedding new light on this family's phylogeny. Shell of juvenile cowries is thin with a wide aperture and a thin outer lip. Shell of an adult cowrie is solid, ovate, pear shaped, flat based with a rounded top and a long, thin aperture that is offset to the right. Unlike the mantle in most gastropods, the mantle of cowries covers the whole shell and gives a glossy appearance. Cypraeids are typically herbivorous. Foot is wide and flat and, when fully extended, may protrude from both ends. On the Texas coast Cypraeidae is represented by 3 genera and 4 species whose size ranges from 12 to 190 mm (½ to 7½ in); 1 species is discussed here.

Macrocypraea cervus (Linnaeus, 1771)

Atlantic Deer Cowrie

Distribution: North Carolina to Florida, Texas; Cuba; Bermuda. **Size:** 76 to 190 mm (3 to 7 in).

Description: Color brown with beige-tan dorsal spots; teeth chocolate brown. Shell shape elongate-oval. Spire concealed with a thick dorsal line. Aperture long and narrow with columellar lip turned inward. Teeth most pronounced on outer labial process, less pronounced on columellar lip. **Habitat:** Common in calcareous environments, such as coral reefs. In Texas found at Flower Garden Banks and Seven and One-half Fathom Reef at depths from 14 to 53 m (45 to 174 ft). Depth range 0 to 53 m (174 ft). **Remarks:** This is the largest cowrie in the western Atlantic. **Synonym:** *Cypraea cervus* Linnaeus, 1771.

Ovulidae

Ovulidae is a large family of gastropods of about 150 species. The shell is pear shaped, elongate, and cylindrical with long anterior and posterior canals that are drawn outward. Sculpture is smooth. The specialized habitat and association with cnidarians distinguish the Ovulidae. Juvenile ovulid shells are thin with a sharp-edged outer lip. Surprisingly, the anatomy of the ovulids is not well known, and what is known shows an anatomy similar to that of cypraeids. In most ovulids the mantle is brightly colored and typically contains distinct patterns that tend to mimic the cnidarian host. Ovulids are carnivorous gastropods, feeding on cnidarian polyps. Little is known of their egg-laying capabilities, but it appears the females lay a clutch of eggs on the surface of the host cnidarian. On the Texas coast Ovulidae is represented by 4 genera and 6 species whose size ranges from 12 to 57 mm (½ to 2¼ in); 3 species are discussed here.

Pseudocyphoma intermedium (Sowerby I, 1828)   **Intermediate Cyphoma**

Distribution: Florida, Texas; Caribbean; Brazil; Bermuda.
Size: 30 mm (1⅕ in).

Description: Color yellowish-white, polished. Shell shape ovate, oblong, tapers at both ends and sharper on anterior end than posterior end. Sculpture smooth with irregular commarginal growth threads and 2 transverse ridges, one located anteriorly and the other slightly posterior of midregion. Aperture somewhat narrow, broader at anterior end, narrowest at infolded posterior end, almost as long as shell. Outer apertural lip thickened and curved inward. Columella smooth without teeth and with single, somewhat obscure plait. **Habitat:** Epifaunal on soft corals at depths from 0 to 128 m (420 ft). **Remarks:** In Texas this shell is located from Port Aransas and southward. Formerly known as _Cyphoma intermedium._

Simnialena marferula Cate, 1973   **Sea-whip Simnia**

Distribution: Texas.
Size: 20 mm (⅘ in).

Description: Color yellowish-orange with parietal area and margin of aperture whitish. Shell shape broadly elongate and tapers on both ends. Sculpture smooth with many irregular, incised spiral lines crossed by few irregular axial growth lines; outer region of anterior canal with fine spiral threads. Aperture broad, becoming wider anteriorly, more so than in one-tooth simnia. Outer lip strong and thickened. Posterior end of columella pinched in with a spiral, toothlike structure. **Habitat:** Attached to sea whip coral (_Leptogorgia setacea_). **Remarks:** Considered by some authors to be a synonym of _S. uniplicata._ Common in beach drift on Texas coast.

Simnialena uniplicata (Sowerby II, 1849)   One-tooth Simnia

Distribution: Virginia to east and west coasts of Florida, Texas; Brazil.
Size: 12 to 19 mm (½ to ¾ in).

Description: Color orangish-brown with outer lip and tooth white. Shell shape elongate, narrow, and tapers at both ends. Sculpture smooth with irregular, spiral lines anteriorly and posteriorly. Aperture almost as long as the shell. Posterior end pinched in with a single spiral, toothlike structure. Outer lip round and thick. **Habitat:** Attached to the gorgonian coral (_Eugorgia virgulata_). Depth range 1 to 116 m (3 to 380 ft). **Remarks:** Occasionally found in beach drift from Port Aransas and southward. Similar to sea-whip simnia. However, sea-whip simnia is generally broader and larger and has fine spiral striations throughout the dorsal surface. One-tooth simnia has spiral striae posteriorly and anteriorly.

Naticidae

Naticidae is a large family of gastropods, also known as moonsnails, found throughout the world. Naticids are cosmopolitan, with the greatest diversity of species found in tropical climates. They have smooth, globose to pear-shaped shells. Shells of naticids are right-handed and medium to low spired with the exception of the genus _Sinum,_ which is nearly ear shaped. Naticids are predatory animals, usually on other mollusks. The molluscan prey is captured by wrapping the prey in the naticid's foot and using its radula in combination with the acid-secreting proboscideal accessory boring organ to drill a hole into the prey shell, from which the flesh is consumed. Naticids are found in sandy or muddy areas of the intertidal and sublittoral zones. In Texas Naticidae is represented by 4 genera and 9 species whose size ranges from 6 to 64 mm (¼ to 2½ in); 4 species are discussed here.

Neverita delessertiana (Récluz, 1843)

Distribution: Texas, Gulf of Mexico.

Size: 27 to 56 mm (1 to 2⅕ in).

Description: Color light tan to yellowish-brown; upper whorls and protoconch darker with a reddish-brown band below the suture; base white. Shell shape semiglobular. Sculpture smooth with irregular, wavy axial lines on teleoconch whorls that continue onto base and into umbilical area. Spire elevated. Aperture broadly ovate. Umbilicus deep with a deep brownish, triangular callus that goes over the umbilical area, attaches to the body whorl, and covers about ⅔ of umbilicus; upper edge of umbilical wall with a ridge. **Habitat:** Sandy shores and bays. Typically found in shallower water than the shark eye. **Remarks:** Not a synonym of _N. duplicata_.

Neverita duplicata (Say, 1822)   **Shark Eye**

Distribution: Cape Ann, Massachusetts, to Florida to Texas.

Size: 25 to 63 mm (1 to 2½ in).

Description: Color grayish to light brown, with older whorls shaded darker. Shell shape globose. Sculpture smooth with fine, radial growth lines. Spire slightly elevated. Aperture ovate and outer lip thin. Umbilicus deep, almost entirely closed by a thick callus; upper edge of umbilical wall without a ridge. **Habitat:** Shallow water of bays, estuaries, and Gulf of Mexico. On Texas coast found at depths from shoreline to about 47 m (158 ft). Found in deeper water than _N. delessertiana_. Depth range 0 to 58 m (190 ft). **Remarks:** Active predator on other mollusks. Commonly located on Texas coast. **Synonym:** _Polinices duplicata_ (Say, 1822).

Sinum perspectivum (Say, 1831) **White Baby Ear**

Distribution: Maryland to Florida to Texas; Caribbean; Bermuda.

Size: 25 to 51 mm (1 to 2 in).

Description: Color uniformly white with a yellowish periostracum, sometimes splotched with brown exteriorly; interior polished white. Shell shape like an ear and extremely flat. Sculpture of numerous paired fine growth lines, with 3 whorls. Apex at same level of body whorl with slightly impressed suture. Aperture wide and roundish with outer lip sharp and thin. **Habitat:** In sand along outer beaches and inlet areas; common in beach drift along the Texas shore. Depth range 0 to 70 m (230 ft). **Remarks:** At low tide, this animal has been observed making a wide track. **Synonym:** _Sigaretus perspectivus_ Say, 1831.

Tectonatica pusilla (Say, 1822) **Miniature Moonsnail**

Distribution: Maine to Florida to Texas to Brazil.

Size: 6 to 8 mm (¼ to ⅓ in).

Description: Color grayish; band near suture and base with reddish-brown flammules; body whorl with broad band of brown and gray maculations. Shell shape globose and ovate, not flattened. Sculpture smooth, porcelaneous, and with fine axial growth lines. Body whorl expanded. Aperture broad, semilunar with a thin outer lip. Umbilicus almost entirely covered by a callus, but with a small opening. **Habitat:** Inlet areas. _T. pusilla_ is usually collected alive from shore to a depth of about 42 m (138 ft). Shells have been collected at depths from 0 to 130 m (426 ft). **Remarks:** _T. pusilla_ is often confused with young _Neverita duplicata;_ however, the operculum of _N. duplicata_ is corneous. One of the most common naticids found on the Texas coast. **Synonym:** _Natica pusilla_ Say, 1822.

Cassidae

Cassidae is a small family of gastropods, also known as helmet shells, of about 60 species found in warm waters worldwide. Cassids are known for their solid globose shells, large body whorl, and low conic spire. The anterior siphonal canal is typically short and narrow and, in most taxa, highly twisted. Many cassids occur intertidally in the sand, with a few preferring hard substrates, but some are found in deeper water of the continental shelf. Feeding studies have shown the cassids to be predatory on echinoids. The cassids produce an acid that removes spines from an area of the echinoderm test prior to drilling a hole in the test and feeding. On the Texas coast Cassidae is represented by 5 genera and 6 species whose size ranges from 25 to 229 mm (1 to 9 in); 1 species is discussed here.

Semicassis granulata (Born, 1778)   **Scotch Bonnet**

Distribution: North Carolina to Texas; Brazil; Bermuda.
Size: 38 to 120 mm (1½ to 5 in).

Description: Color of base creamish-tan with spiral rows of reddish-brown, squarish spots. Shell shape broadly oval with an extended spire. Sculpture of axial grooves and spiral cords, giving beaded checkerboard appearance that is most pronounced on spire. Body whorl approximately ¾ shell height; lower parietal region pustulose. Aperture half-moon shaped and length of body whorl, with a thick outer lip that has teeth on both sides. Siphonal canal turned upward dorsally. **Habitat:** Sandy bottoms from intertidal zone to a depth of approximately 183 m (600 ft). **Remarks:** State Shell of North Carolina. **Synonym:** *Phalium granulata* (Born, 1778).

Personidae

Personidae is a small family of gastropods that are closely related in shell characters. The dominant personid genus, *Distorsio,* is exceptional because of its asymmetrical coiling, which becomes more distinct as the animal matures. Some authors believe that apertural features remain constant as the animal grows in order to house the head-foot region. This causes the asymmetrical coiling, which is greater in diameter than the aperture itself. The densely armored aperture presumably offers protection from predators as well as ease of movement into and out of the shell. There is little information on personid ecology and reproduction, but the wide distribution suggests that most species have planktotrophic larval stages. On the Texas coast Personidae is represented by 1 genus and 2 species whose adult size ranges from 18 to 89 mm (¾ to 3½ in); only the most common species is discussed here.

Distorsio clathrata (Lamarck, 1816) — Atlantic Distorsio

Distribution: North Carolina to Texas; Caribbean; Brazil.
Size: 18 to 89 mm (¾ to 3½ in).

Description: Color dull white; parietal shield may be stained orangish-brown. Shell shape ovate-conic. Sculpture densely latticed with spiral and axial ridges regularly spaced and distinct, producing knobs at intersection; varices short. Spire somewhat high. Body and penultimate whorls occurring at different angles appear distorted. Aperture also distorted, strongly toothed, glossy white, with numerous denticles and distinct, irregularly shaped siphonal canal. **Habitat:** On sand and under coral just below waterline to a depth of 300 m (984 ft). **Remarks:** Similar to *D. constricta mcgintyi*. However, *D. constricta mcgintyi* has only 1 parietal tooth, and *D. clathrata* has 2.

Ranellidae

Ranellidae is a relatively large family of gastropods, also known as the triton shells. All ranellids exhibit a fusiform shell shape; however, shells of ranellids vary in height of apex and length of siphonal canal. Sculpture, habitats, and geographic range also vary for this family. Ranellids are moderately large, intertidal to shelf epifaunal snails that inhabit rocky shores, coral reefs, and sandy substrates. All ranellids are carnivores and typically feed on asteroids, echinoids, tunicates, and various types of carrion. Ranellids are some of the most widely distributed mollusks, which in large part is due to their long, free-swimming veliger stage (approximately 3 months), when they are able to feed on plankton, allowing the veligers to hold off metamorphosis until they locate a suitable substrate. Economically, these mollusks have become pests in the mariculture bivalve industry. On the Texas coast Ranellidae is represented by 3 genera and 12 species whose size ranges from 50 to 350 mm (2 to 14 in); 1 species is discussed here.

Cymatium parthenopeum (von Salis, 1793)   **Giant Triton**

Distribution: North Carolina to Texas to Brazil; Bermuda.
Size: 100 to 170 mm (4 to 7 in).

Description: Yellowish to brown with brown and white cords. Shell shape fusiform. Sculpture of spiral cords crossed by numerous axial ridges with dense, fine axial threads in between axial ridges. Possesses 7 to 8 teleoconch whorls; shoulders of the whorls convex. Adult snails have 2 low varices. Spire high and extended. Aperture broadly subelliptical; outer lip with numerous paired teeth; and alternating bands of brown and white. Numerous toothlike structures on parietal wall. Siphon somewhat abrupt. Anal canal bordered by a strong ridge. **Habitat:** On sand, mud, and rubble and under rocks. In Texas found at Stetson Bank and Seven and One-half Fathom Reef. Depth range from 0 to 75 m (246 ft). **Remarks:** Rare species in Texas.

Tonnidae

Tonnidae, known as tun or cask shells, is a small family of gastropods made up of about 20 species. Tonnids are found worldwide in all tropical marine waters. These broad and somewhat oval-shaped snails are medium to large in size. Their shells are typically thin, fragile, and subspherical. The spire is somewhat low. Sculpture is usually of spiral cords and a wide, open aperture. The apertural outer lip is narrowly thickened and flared. The anterior siphonal canal is usually short and widely open. Tonnids typically live in shallow to moderate depths on the continental shelf and slope. Their habitat is soft substrate, where they presumably bury themselves when not feeding. Tonnids are carnivores, and their feeding apparatus allows for the ingestion of whole prey, typically sea cucumbers. In Texas Tonnidae is represented by 1 genus and 2 species whose size ranges from 50 to 178 mm (2 to 7 in); only the most common species is discussed here.

Tonna galea (Linnaeus, 1758) **Giant Tun**

Distribution: North Carolina to Texas; Caribbean; Brazil. **Size:** 127 to 178 mm (5 to 7 in).

Description: Color uniformly white to brown, sometimes with mottlings of darker brown; periostracum brownish. Shell shape globose. Sculpture of spiral ribs intersected by fine axial growth lines. Shell with protoconch made of 2½ smooth whorls. Body whorl dominates shell. Aperture subovate, large, with a weak crenulated outer lip and large parietal shield. Columella short and twisted with a ridge along the outer edge, ending in an abrupt siphonal canal. **Habitat:** Sandy bottoms at depths from 0 to 2359 m (7738 ft). **Remarks:** Found along entire Texas coast, but live material is rare.

Epitoniidae

Epitoniidae is a large family of gastropods made up of hundreds of species found throughout the world in all seas. Shells are usually tall and conic with convex whorls and pointed spires. The first postnuclear whorls are often slightly dorso-ventrally compressed with variable sculpture. Teleoconch whorls are typically circular and convex; however, some species are flat sided, which usually produces dominant axial sculpture. Epitoniids are carnivores that typically feed on cnidarians. These gastropods are benthic dwellers that range from intertidal waters to abyssal depths. On the Texas coast Epitoniidae is represented by 6 genera and 23 species whose size ranges from 5 to 63 mm (⅕ to 2½ in); 9 species are discussed here.

Amaea mitchelli (Dall, 1896)   **Mitchell's Wentletrap**

Distribution: Texas to Panama.

Size: 38 to 63 mm (1½ to 2½ in).

Description: Color whitish-yellow with a broad brownish band in the middle of each teleoconch whorl that extends onto base of shell. Shell shape elongate-conic. Sculpture of low axial and spiral costae of equal size and strength, producing a cancellate appearance. Aperture subcircular and with a thickened columella. Umbilicus not present. **Habitat:** Offshore and occasionally washed ashore. Depth range 0 to 55 m (180 ft). **Remarks:** This species is highly sought after by Texas shell collectors and is often associated with sea anemones. The Coastal Bend Shell Club of Corpus Christi uses this beautiful shell as its logo.

Epitonium albidum (d'Orbigny, 1842)   **Bladed Wentletrap**

Distribution: South Florida to Brazil; Texas; Bermuda; West Africa.

Size: 12 to 20 mm (½ to ⅘ in).

Description: Color glassy white. Shell shape broad and conic. Sculpture of axial, bladelike structures, which are relatively low and appear fused with costae from the previous whorl. Teleoconch with rounded convex whorls and body whorl having about 12 to 14 bladelike costae. Protoconch smooth. Aperture subcircular with the outer lip expanded and flared backward anteriorly. Umbilicus not present. **Habitat:** Beach drift to offshore. Depth range from 0 to 366 m (1200 ft). **Remarks:** Live specimens are found in relatively deep water. In the Bahamas this snail is found associated with a sea anemone. **Synonyms:** *E. ligatum* C. B. Adams, 1850; *E. quindecimcostatum* Mörch, 1874.

Epitonium angulatum (Say, 1831) **Angulate Wentletrap**

Distribution: New York to Florida to Texas; Bermuda.

Size: 18 to 25 mm (¾ to 1 in).

Description: Color white. Shell shape elongate-conic. Sculpture of bladelike axial ridges forming angles on each shoulder that are stronger on earlier whorls. Axial ridges fused with axial ridges from subsequent whorls. Aperture subcircular and with a thickened outer lip held away from body whorl by a carina. Umbilicus not present. **Habitat:** In rubble or sandy bottoms, usually in association with sea anemones. Shells found at depths from 0 to 219 m (718 ft). **Remarks:** Specimen in photograph from Port Isabel.

Epitonium apiculatum (Dall, 1889) **Semismooth Wentletrap**

Distribution: North Carolina, Texas; Puerto Rico.

Size: 5 mm (⅕ in).

Description: Color white. Shell shape elongate-conic. Sculpture of convex whorls, attached by bladelike ridges. Ridges connected to the ridges of subsequent whorls. Protoconch smooth, towerlike, tapering, and with spiral threads on only the first 3 whorls following the protoconch. Aperture roundish with a thickened lip. Umbilicus not present. **Habitat:** Its rare occurrence found alive in beach drift would indicate its habitat to be along the shore. Shells collected at depths from 0 to 90 m (295 ft). **Remarks:** Some authors suggest this species to be an aberrant form of another _Epitonium._

Epitonium humphreysii (Kiener, 1838) — Humphrey's Wentletrap

Distribution: Cape Cod, Massachusetts, to Florida to Texas.
Size: 12 to 25 mm (½ to 1 in).

Description: Color white. Shell shape elongate-conic. Sculpture of strong, numerous, blade-like costae with 9 to 10 convex whorls attached to one another by the costae. Earlier whorls' costae more hooklike and bladelike; latter whorls' costae rounded and thickened. Inner spaces smooth. Aperture subcircular and with an expanded and thick outer lip. Umbilicus not present. **Habitat:** Common from shoreline to a depth of approximately 95 m (312 ft). **Remarks:** Very similar to *E. angulatum;* however, *E. humphreysii* is narrower posteriorly, and the costae are generally broader and rounder than in *E. angulatum.*

Epitonium multistriatum (Say, 1826) — Many-ribbed Wentletrap

Distribution: Massachusetts to Florida to Texas; Bermuda.
Size: 12 to 15 mm (½ in to ⅗ in).

Description: Color white. Shell shape conic. Sculpture of 8 to 10 convex whorls, with latter whorls unattached. Suture well developed and distinct. Axial sculpture with numerous cords to low, bladelike costae. Spiral sculpture of fine striae that do not cross the costae and are light in structure. Aperture subcircular to ovate with a slightly thickened, narrowly expanded lip. Umbilicus not present. **Habitat:** Shells common in beach drift. Depth range 0 to 219 m (720 ft). **Remarks:** The numerous costae on the whorls, particularly on the early whorls, distinguish this species from other *Epitonium* species.

Epitonium novangliae (Couthouy, 1838)   **New England Wentletrap**

Distribution: Virginia to Texas to Brazil; Bermuda.
Size: 12 to 20 mm (½ to ⅘ in).

Description: Color uniformly white to light brown. Shell shape elongate and attenuate. Sculpture of axial blade to cordlike costae. Intercostae made up of spiral threads. Whorls of spire attached. Has 8 to 10 convex whorls separated by a distinct suture and attached by costae only. Aperture subcircular, with outer lip narrowly expanded. Umbilicus narrow and hidden by the parietal lip. **Habitat:** From fairly shallow water to a depth of 457 m (1500 ft). **Remarks:** Fresh specimens mottled brown, but this coloration is typically absent, leaving a white to brownish appearance.

Epitonium rupicola (Kurtz, 1860)   **Brown Band Wentletrap**

Distribution: Cape Cod, Massachusetts, to Florida to Texas.
Size: 12 to 25 mm (½ to 1 in).

Description: Color yellowish-white with a light spiral band below the suture, a dark brown spiral band below the light band, and another dark brown spiral band below the light band and above the succeeding suture. Shell shape elongate-conic. Sculpture of bladelike to rounded axial costae, with 11 globose, convex, attached whorls. Mature specimens typically with 1 to 6 varices. Single spiral thread at base body whorl. Aperture subcircular with outer lip reflected and thickened. **Habitat:** Sandy bottoms at depths from 0 to 65 m (213 ft). **Remarks:** Forages for sea anemones.

Epitonium sericifila (Dall, 1889)   **Silky Wentletrap**

Distribution: Texas; Honduras.

Size: 6 to 10 mm (¼ to ⅖ in).

Description: Color white. Shell shape slenderly conic. Sculpture of 10 whorls attached and angled at the periphery. Axial sculpture of numerous, low, diagonally angled, somewhat connected rounded costae that extend to base of body whorl. Spiral sculpture of numerous fine spiral threads in between costae but do not cross costae. Protoconch smooth and tapered. Aperture subcircular. **Habitat:** Shallow inlet areas at depths from 0 to 7 m (23 ft). **Remarks:** The sculpture of this exquisite epitoniid gives it a sheen or glistening texture.

Janthinidae

Janthinidae, commonly known as the purple or violet sea snail family, is a small family of pelagic gastropods of approximately 8 species. Janthinids are common in tropical and warm temperate seas. The shells are thin and fragile, globose, and somewhat top shaped; most are violet in color. The shell is moderate to small in size and composed almost entirely of calcite, making it somewhat sturdy but lightweight on the water. The shell is typically smooth but may have fine axial and/or spiral growth lines. Most construct a mucus float by means of an air-bubble raft with the foot uppermost, facing the air/water interface, which affords janthinids the ability to sustain themselves at the surface of the water. They are found in shoals or in groups on the surface of the open ocean with some shoals or aggregates being recorded at about 370 km (200 nautical mi) across. These mollusks are carnivores. On the Texas coast Janthinidae is represented by 2 genera and 5 species whose size ranges from 6 to 38 mm (¼ to 1½ in); 3 species are discussed here.

Janthina globosa Swainson, 1822   Elongate Janthina

Distribution: Texas; shores of US east and west coasts; Bermuda. **Size:** 12 to 18 mm (½ to ¾ in).

Description: Color purple with body whorl a paler purple. Shell shape globose. Sculpture smooth with faint axial growth lines. Protoconch small, bead-like, with a portion immersed into succeeding whorl. Body whorl broad, extended, and round. Aperture with extended outer lip acutely flared posteriorly and anteriorly; at the depression of the aperture are growth lines that curve backward, forming a keel. Outer lip thin and depressed or notched anteriorly. **Habitat:** Circumtropical; pelagic in warm seas at depths from 0 to 13 m (43 ft). **Remarks:** Shells are not common in beach drift.

Janthina janthina (Linnaeus, 1758)   Janthina

Distribution: Texas; US east and west coasts; Brazil; Bermuda. **Size:** 25 to 38 mm (1 to 1½ in).

Description: Color paler purple on top of shell, deeper purple at base. Shell shape low and broad; body whorl large, angular, and compressed. Sculpture smooth with irregular, faint axial and spiral growth lines; spiral lines thickened, more pronounced, and at more regular intervals at base. Aperture somewhat broad, subquadrate with a thin outer lip. Anal canal somewhat angled, and parietal shield curls backward. **Habitat:** Circumtropical; pelagic in warm seas. **Remarks:** Gulf Stream carries these snails as far north as Massachusetts. Purple coloration may be for camouflage from fish and birds.

Janthina pallida (Thompson, 1840)   **Pale Janthina**

Distribution: Pelagic; worldwide in warm seas.
Size: 12 to 18 mm (½ to ¾ in).

Description: Color purplish-white with lavender on outer lip; interior violet. Shell shape globose. Sculpture smooth with fine, irregular axial and spiral growth lines. Body whorl swollen and broadly round. Aperture subquadrate, D-shaped, and flaring at base and upper angle of parietal lip; anterior end of aperture flared, and posterior portion somewhat rounded. **Habitat:** Circumtropical; pelagic in warm seas. **Remarks:** Apex or spire somewhat depressed and lower than in _J. globosa._

Eulimidae

Eulimidae is a family of about 8 genera and hundreds of species of typically parasitic marine gastropods found in all seas. Shell varies from high spired long, narrow, and curved to inflated globose. The shell is usually small (less than 5 mm, or ⅕ in) and solid. The surface is usually smooth, but some species may have distinct sculpture. Most primitive shells have a straight, conic shell with flat-sided whorls, smooth surface, and high spire. Most eulimids are ectoparasites permanently attached to their host by the snout or proboscis. Many eulimids show sexual dimorphism, with males being anywhere between 0.1 and 0.7 times smaller than females. On the Texas coast Eulimidae is represented by 7 genera and 12 species whose size ranges from 3 to 14 mm (⅛ to ½ in); 6 species are discussed here.

Eulima bifasciata d'Orbigny, 1841 **Two-band Eulima**

Distribution: Florida, Texas; Caribbean.
Size: 7 mm (⅓ in).

Description: Color translucent light tan, almost white, polished with 2 brown bands on each teleoconch whorl, one just below the suture and the other just above the preceding whorl. Shell shape mildly vertical and slender. Sculpture smooth and polished. Suture indistinct with exception of colored bands. Protoconch made up of 2 smooth whorls, with first whorl semilunar. Aperture teardrop shaped with outer lip relatively straight, tall, and thin. **Habitat:** Outer Texas beaches from shoreline to a depth of 84 m (275 ft). **Remarks:** Rarely found in bays. **Synonym:** _Strombiformis bifasciatus_ (d'Orbigny, 1842).

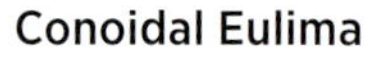

Melanella conoidea (Kurtz and Stimpson, 1851) **Conoidal Eulima**

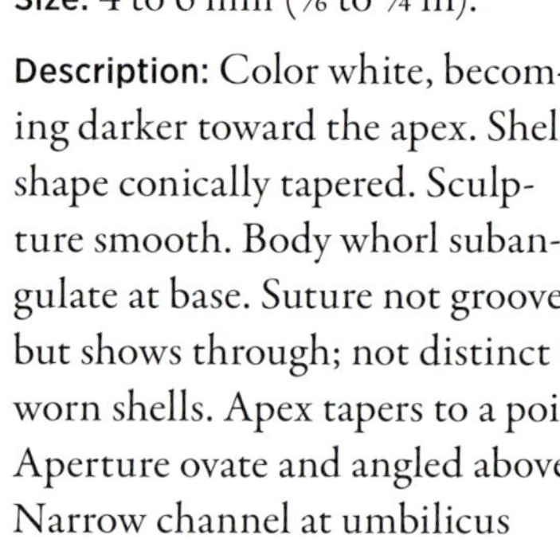

Distribution: Florida, Texas; Caribbean.
Size: 4 to 6 mm (⅙ to ¼ in).

Description: Color white, becoming darker toward the apex. Shell shape conically tapered. Sculpture smooth. Body whorl subangulate at base. Suture not grooved but shows through; not distinct in worn shells. Apex tapers to a point. Aperture ovate and angled above. Narrow channel at umbilicus separated by a keeled inner parietal lip. **Habitat:** In rubble and sandy bottoms offshore; rare in bays. Commonly found off Galveston and Freeport at depths from 14 to 38 m (48 to 126 ft). Depth range 0 to 538 m (1765 ft). **Remarks:** Ectoparasite in relatively shallow water. Separated from _M. sarsi_ by its somewhat compact base, keeled body whorl, and pointed spire.

Melanella jamaicensis (C. B. Adams, 1845)   Jamaica Eulima

Distribution: Virginia to Florida, Texas; Caribbean.

Size: 6 to 9 mm (¼ to ⅔ in).

Description: Color of spire and base glossy white; area in between sometimes brown. Shell shape slender and conic. Sculpture of smooth, flat-sided whorls with ill-defined sutures; whorls taper to a sharp apex. Aperture teardrop shaped and slender. Outer lip thin and slightly flared at base. Umbilicus made up of a short, narrow channel behind a somewhat thickened columella. **Habitat:** Ectoparasite on sea cucumbers; in bays. The shell is commonly found in beach drift. Depth range 0 to 366 m (1200 ft). **Remarks:** Probably the most common eulimid on the Texas coast in beach drift. Live material can be found on the shelf area off the Texas coast.

Microeulima hemphillii (Dall, 1884)   **Brown Eulima**

Distribution: Florida, Texas; Brazil.

Size: 3 mm (⅛ in).

Description: Color translucent tan with a narrow band of brown above the suture. Shell shape slender and conic. Sculpture smooth. First whorl on protoconch swollen. Shell has approximately 8 flat-sided whorls with a well-rounded body whorl. Suture defined by a narrow thread. Aperture teardrop shaped, not as vertical as in *E. bifasciata*. **Habitat:** All along Texas coast, including bays and offshore dredging. Depth range 0 to 48 m (157 ft). **Remarks:** *M. hemphillii* in Texas typically does not reach the size of Florida *M. hemphillii*. The largest recorded specimen for Texas is 2.8 mm (⅛ in). **Synonym:** *Eulima hemphillii* Dall, 1884.

Niso aeglees Bush, 1885 — **Brown-line Niso**

Distribution: North Carolina to Texas; Caribbean; Brazil.
Size: 10 to 14 mm (⅖ to ⅗ in).

Description: Color pale brown with mottlings of darker brown; brown line at base of suture. Shell shape conic, tapered with an acute spire. Sculpture smooth and with a body whorl broad and angled at the base with flat-sided whorls. Aperture pear shaped with a sharp outer lip. Umbilicus deep and chinklike behind the thickened columella. **Habitat:** Found on shelf area of Texas-Louisiana coast and in shallow bays. Depth range from 0 to 260 m (853 ft). **Remarks:** Andrews (1977) reports this species from bays and inlet-influenced locations. However, Odé (1989) states this species does not enter bay areas. **Synonym:** *N. interrupta* auct. non Sowerby, 1834.

Vitreolina arcuata (C. B. Adams, 1850) — **Twisted Eulima**

Distribution: North Carolina, Texas; Caribbean.
Size: 4 mm (⅙ in).

Description: Color translucent white. Shell shape curved and conic. Sculpture smooth with convex whorls, lightly impressed sutures, and a fine spiral line above the suture. Apex curved significantly in upper whorls. Aperture ovate to teardrop shaped. **Habitat:** Commonly found on the beaches of South Padre Island, Matagorda, and Galveston. It has also been found in dredgings offshore Galveston and Freeport and in sediment samples from Stetson Bank and Flower Garden Banks. Depth range from 0 to 166 m (545 ft). **Remarks:** This snail is known to be ectoparasitic on sea cucumbers. **Synonym:** *Eulima arcuata* C. B. Adams, 1850.

Cerithiopsidae

Cerithiopsidae is a large family of gastropods of about 150 small (less than 10 mm, or ⅖ in) species. Shells are typically tall, narrow, many whorled, and cylindrical with variable sculpture. These gastropods are usually found living on sponges at all depths. For taxonomic purposes, the variation in sculpture and shape of the protoconch are important. Cerithiopsids feed on the sponges on which they live, like the triphorids. On the Texas coast Cerithiopsidae is represented by 3 genera and 15 species whose size ranges from 3 to 12 mm (⅛ to ½ in); the 3 most common shallow-water species are discussed here.

Cerithiopsis greenii (C. B. Adams, 1839)   Green's Miniature Cerith

Distribution: Cape Cod, Massachusetts to east and west coasts of Florida, Texas; Bermuda; Brazil. **Size:** 3 mm (⅛ in).

Description: Primary color dark chocolate brown, with darkest shade of brown at base and protoconch a translucent lighter brown. Shell barrel shaped and narrowly fusiform. Protoconch smooth. Whorl preceding body whorl with 2 glassy beaded cords; latter whorls with 3 beaded cords on each whorl; beaded cords attached by spiral and axial threads that produce inner spaces. Aperture somewhat oval and with abrupt anal and siphonal canals. **Habitat:** In bays, inlets, and shelly sand at depths from 0 to 75 m (246 ft). **Remarks:** Often confused with *C. gemmulosa* and *C. fusiformis.* However, *C. greenii* is smaller and darker and has a higher spire than *C. gemmulosa;* and *C. fusiformis* has a broader spire, is lighter brown than *C. greenii,* and is darker than *C. gemmulosa.*

Retilaskeya bicolor (C. B. Adams, 1845) **Awl Miniature Cerith**

Distribution: Massachusetts to Caribbean; Texas; Brazil. **Size:** 12 to 18 mm (½ to ⅔ in).

Description: Color of teleoconch uniformly brown and protoconch light brown. Shell shape elongate-conic and slender. Shell of 10 to 14 whorls. Teleoconch whorls composed of spiral threads crossed by axial ribs, producing a beaded latticelike appearance with 2 spiral beaded ribs on whorls succeeding the protoconch and 3 on subsequent whorls. Shell suture distinct. Aperture subcircular and with a thin outer lip. Siphonal canal short. **Habitat:** In shelly sand along shoreline to depths of approximately 65 m (213 ft). **Remarks:** Beads on spiral cords of the genus *Retilaskeya* are smaller than on *Cerithiopsis.* Complete specimens with protoconchs are rare. **Synonym:** Mistakenly called *C. emersoni* C. B. Adams, 1839.

Seila adamsii (Lea, 1845) **Adam's Miniature Cerith**

Distribution: Massachusetts to Florida, Texas to Brazil; Bermuda. **Size:** 6 to 12 mm (¼ to 1/ in).

Description: Color uniformly brownish-yellow to dark brown. Shell shape turriculate and narrowly conic. Sculpture of 3 squarish spiral cords per whorl with fine axial threads between each cord, and with a somewhat indistinct suture between whorls. Protoconch abrupt and bulbous. Aperture subquadrate with outer portion of lip crenulate. Siphonal canal short with 2 somewhat pointed ends. **Habitat:** Hypersaline bays and shallow water to a depth of 104 m (340 ft). This snail is typically found under shell and on sponges and rocks. **Remarks:** Whorls not easily distinguished. However, species is easily identified.

Triphoridae

Triphoridae is a large family of gastropods with hundreds of species found in tropical and warm temperate waters worldwide. Triphorids are the only group of marine gastropods that are predominantly sinistral (left-handed). One exception is the dextral (right-handed) genus *Metaxia*. Triphorid shells are typically small (less than 10 mm, or ⅖ in). Triphorids normally feed on sponges by using an extendable proboscis equipped with a modified esophagus for adding length. Triphorids are common under rocks and dead coral slabs in shallow water; however, some may be found as deep as 1000 m (3300 ft). In Texas Triphoridae is represented by 8 genera and 10 species whose size ranges from 5 to 9 mm (⅕ to ⅔ in); 2 species are discussed here.

Cosmotriphora melanura (C. B. Adams, 1850)   **White Atlantic Triphora**

Distribution: North Carolina to Florida to Brazil; Texas.
Size: 5 to 8 mm (⅕ to ⅓ in).

Description: Color snow white with a dark brown protoconch. Shell shape tall, slender, and conic. Sculpture of beaded spiral cords connected by axial rows of rounded beaded ribs. Axial ribs and spiral beaded rows continue to the base of the shell. The early teleoconch whorls have only 2 rows of beaded cords, and latter whorls have 3. Suture dividing the whorls somewhat distinct. Protoconch of approximately 4½ whorls composed of axial ribs crossed by 2 raised spiral threads. Aperture roughly subcircular, and siphonal canal distinctly abrupt. **Habitat:** Tropical calcareous environments at depths from below shoreline to 91 m (300 ft). In Texas this snail is found on offshore reefs and banks. **Remarks:** This is an exquisitely beautiful shell. **Synonym:** *Triphora melanura* C. B. Adams, 1850.

Marshallora nigrocincta (C. B. Adams, 1839) **Black-lined Triphora**

Distribution: Massachusetts to Florida to Texas to Brazil; Bermuda.
Size: 5 to 10 mm (⅕ to ⅖ in).

Description: Color brown with lighter beads and a blackish-brown band at the suture; protoconch, whorls following protoconch, and base of the body whorl dark brown. Shell shape broad and conical. Sculpture of 10 to 12 convex whorls that taper to an acute apex. Teleoconch whorls with 2 beaded spiral cords attached by axial, rounded riblets, with the exception of the body whorl, which has 3 beaded spiral cords. Suture incised, distinct, and as dark brown as the keeled base that is composed of strong revolving cords. Aperture oval and oblique with a well-formed short siphonal canal. **Habitat:** Found on seaweed at low tide in high-salinity areas at depths from 1 to 70 m (3 to 230 ft). **Remarks:** Recorded as a synonym of *M. modesta* (C. B. Adams, 1850). **Synonym:** *Cerithium nigrocinctum* C. B. Adams, 1839.

Muricidae

Muricidae is a large family of marine predatory snails made up of approximately 100 genera and about 700 species found globally. Members of this family exhibit high variability. Shells may be limpet shaped to spindle shaped but rarely smooth. Sculpture is usually highly ornamented and is what makes this family one of the most beautiful. Many of the numerous genera found in the muricid family are ill-defined, and confusion exists as to their taxonomic status. Muricids are found in shallow tropical and subtropical waters, on rocks and reefs, with fewer species in soft substrate. Muricids are carnivores that feed on a variety of invertebrates such as other gastropods, bivalves, corals, barnacles, and polychaete worms. On the Texas coast Muricidae is represented by 14 genera and 26 species whose size ranges from 8 to 178 mm (⅓ to 7 in); the 4 most common shallow-water species are discussed here.

Hexaplex fulvescens (Sowerby II, 1834)   **Giant Eastern Murex**

Distribution: North Carolina, Florida, Texas.
Size: 127 to 185 mm (5 to 7⅓ in).

Description: Color of shell whitish with rust color on portions of axial and spiral elements and spines; periostracum brown, and inner aperture white. Shell shape ovate. Sculpture on whorls heavily spinose. Axial sculpture of strong cords and spinose varices. Spiral sculpture of strong cords that connect spines to varices and have raised threads between cords; spines irregular in shape and size. Suture indistinct and hidden by succeeding whorl. Spire short and pointed. Aperture round and broad. Siphonal canal short and wide, and older siphonal canals appear to be an umbilicus. **Habitat:** In shallow areas off the coast of Texas. Depth range 0 to 79 m (260 ft). **Remarks:** One of the largest muricids in the western Atlantic. This snail is a predator of bivalves. **Synonym:** *Murex fulvescens* Sowerby II, 1834.

Phyllonotus pomum (Gmelin, 1791) **Apple Murex**

Distribution: North Carolina to Florida to Brazil; Texas; Bermuda.

Size: 80 mm (3 in).

Description: Color from brownish-yellow to darker brown with some individuals having different tints of brown bands and dark spots on the outer lip. Shell shape somewhat broad and spindlelike. Sculpture of convex teleoconch whorls with 3 varices on each whorl, with hollow, blunt spines or nodules on the varices. Intervarical axial ribs crossed by irregular, scaly, spiral cords. Aperture subcircular with outer lip dentition nodulose and parietal shield pressed to body whorl. Columella smooth. **Habitat:** Grass beds, sand, and hidden under rocks. Depth range 0 to 73 m (240 ft). **Remarks:** Similar to *P. oculatus,* but parietal lip narrower and inner area of outer lip dentition nodulose. **Synonym:** *Murex pomum* Gmelin, 1791.

Stramonita canaliculata (Gray, 1839) **Hays' Rocksnail**

Distribution: Texas, northwest Florida.

Size: 112 mm (4½ in).

Description: Color grayish with some specimens having a darker color in a spiral or axial pattern; interior of aperture brownish-pink. Shell shape broadly conic. Sculpture of coarse spiral incised threads, with 2 spiral rows of large nodulose ridges at the whorl shoulders. Spire somewhat extended. Body whorl angulate. Aperture subovate with outer lip thick and crenulate and inner lip smooth with thickened parietal shield. Siphonal canal obtuse and short. **Habitat:** Shallow areas on hard substrates such as rocks or oyster reefs at depths from 1 to 11 m (3 to 33 ft). **Remarks:** Specimens with fewer than 6 whorls have oblique tubercles. **Synonyms:** *Thais haemastoma canaliculata* (Gray, 1839); *T. haemastoma haysae* Clench, 1927.

Stramonita haemastoma (Linnaeus, 1767)   Florida Rocksnail

Distribution: North Carolina to Florida, Texas; Caribbean; Brazil.

Size: 51 to 76 mm (2 to 3 in).

Description: Color grayish to dark brown with some specimens having an axially patterned darker color of brown maculations. Shell shape ovate with a raised spire. Sculpture variable, but whorls typically slightly convex and occasionally angled, with spiral rows of incised lines that may have rows of different-sized nodules spirally arranged. Axial sculpture of numerous cords or ribs with irregular growth lines between. Aperture broadly subovate with a thick and crenulate outer lip. Inner lip with smooth parietal shield. Siphonal canal short. **Habitat:** Feeding on bivalves of South Texas jetties and rocky shores. Depth range 0 to 538 m (1765 ft). **Remarks:** A variable species with numerous synonyms. **Synonym:** *Thais haemastoma* (Linnaeus, 1767).

Buccinidae

Buccinidae is a large family of gastropods of about 200 species found from pole to pole in all seas. Buccinids are primarily a marine family with small to medium-sized globose, ovate to fusiform, strong- to weak-shouldered shells. Habitat is variable. Most buccinids inhabit subtidal rubble or sand, and some are adapted to hard substrates. The majority of buccinids are scavengers or predators; however, a few species are limited to specific prey. Buccinids are a poorly understood family. In Texas Buccinidae is represented by 7 genera and 8 species whose size ranges from 18 to 31 mm (¾ to 1¼ in); 2 species are discussed here.

Gemophos tinctus (Conrad, 1846) — Tinted Cantharus

Distribution: North Carolina to Florida to Texas; Caribbean; Bermuda; Brazil.

Size: 18 to 35 mm (¾ to 1½ in).

Description: Color is variable, from blue-gray to yellowish to chocolate to milk white. Shell shape elongate-conic. Sculpture of 5 to 6 convex whorls. Spiral sculpture of cords with finer striae in between. Axial sculpture of somewhat strong axial cords that produce weak nodules where they intersect the spiral sculpture. Aperture ovate with a thickened inner and a crenulate outer lip that has larger teeth on upper part of outer lip, which borders the anal canal. Parietal shield smooth and terminates at an angle. **Habitat:** Sandy, shelly, and rocky habitats from shoreline to a depth of 80 m (262 ft). **Remarks:** Closest relative *G. auritulus.* **Synonym:** *Pisania tinctus* (Conrad, 1846).

Solenosteira cancellaria (Conrad, 1846) — Cancellate Cantharus

Distribution: Florida to Texas; Yucatán.

Size: 18 to 36 mm (¾ to 1¼ in).

Description: Color yellowish to reddish-brown. Shell shape elongate-conic. Sculpture of 5 to 6 convex whorls, with spiral and axial cords that cross and give a beaded, cancellate appearance; fine axial and spiral threads between thicker axial and spiral cords. Aperture ovate with a crenulate outer lip. Siphonal canal narrow, ends abruptly, slightly upturned, and angled posteriorly. Anal canal short, and parietal shield somewhat narrow and smooth. **Habitat:** Typically found on rocks and reefs from shoreline to a depth of 101 m (330 ft). **Remarks:** In Texas found on sand and mud substrate. Shells are commonly found with hermit crabs in bays. **Synonym:** *Cantharus cancellarius* (Conrad, 1846).

Columbellidae

Columbellidae is a large family of gastropods, commonly known as dovesnails, comprising several hundred species found throughout the world. These snails have elongate to ovate shells. The siphonal canal is usually short, and the aperture is typically narrow. Sculpture can be smooth or ornamented, with a variable spire that can be high and pointed to low and broad. The outer and inner lips of the aperture are usually denticulate. Radular features of the Columbellidae have divided this group into 2 subfamilies, Columbellinae (herbivorous) and Pyreninae (carnivorous or omnivorous). In Texas Columbellidae is represented by 8 genera and 13 species whose size ranges from 4 to 24 mm (⅙ to 1 in); the most common 6 species are discussed here.

Astyris lunata (Say, 1826)   Lunar Dovesnail

Distribution: Massachusetts to Florida to Texas to Brazil; Bermuda.

Size: 6 mm (¼ in).

Description: Color variable, typically tannish-white with axially oriented irregular, zigzag brownish to purplish bands and brown apex. Shell shape broadly ovate-conic. Sculpture smooth with convex whorls and impressed sutures. Aperture narrowly ovate; inner area of outer lip thin and smooth, with an arched or distinctly angled columella, and base of body whorl with fine spiral grooves. **Habitat:** Low-tide line on grass and rocks in bays at depths from 0 to 115 m (377 ft). **Remarks:** Similar to *A. multilineata*. However, *A. lunata* is broader basally, the outer lip is not noticeably denticulate, and coloration is not as vibrant. **Synonym:** *Mitrella lunata* Say, 1826.

Costoanachis avara (Say, 1822)   Greedy Dovesnail

Distribution: Massachusetts to Florida to Texas.

Size: 10 to 22 mm (⅖ to ⅘ in).

Description: Color variable from straw yellow to chestnut brown with white elliptical mottlings. Shell shape elongate-conic. Sculpture of axial, swollen, randomly arranged ribs on whorls that fade toward base. Spiral sculpture of microscopically fine threads dorsally that become larger ventrally. Protoconch smooth, and succeeding whorls with axial riblets. Aperture somewhat narrowly elliptical; outer lip curved and thick, and inner surface with toothlike structures. Columella narrow and smooth, and the siphonal canal slightly curved. **Habitat:** Found in grass beds and under stones, in bays, in inlets, and on beaches. At depths from below tide line to 46 m (150 ft). **Remarks:** Do not confuse with *C. sparsa* (Reeve, 1859), whose axial sculpture is more uniform than that of *C. avara*. **Synonym:** *Anachis avara* (Say, 1822).

Costoanachis floridana (Rehder, 1939)   Florida Dovesnail

Distribution: North Carolina to Florida, Texas.

Size: 12 mm (½ in).

Description: Color variable from yellowish to white with irregular axial, spiral, and zigzag-type markings of various tones of reddish-brown. Shell shape elongate-conic. Sculpture variable, typically of axial ribs on upper rounded whorls and upper area of the somewhat rounded and broad body whorl. Spiral sculpture of fine threads on convex whorls, with spiral threads becoming larger toward base. Aperture narrow and long with smooth columella. Inner surface of thickened outer lip denticulate.

Habitat: On rocks and sand. Found along Texas coastal bays and beaches from shoreline to a depth of 14 m (45 ft). **Remarks:** Do not confuse with *C. semiplicata,* whose upper whorls are not as rounded. **Synonym:** *Anachis floridana* Rehder, 1939.

Costoanachis semiplicata (Stearns, 1873) **Semiplicate Dovesnail**

Distribution: Florida, Texas; Yucatán.

Size: 15 mm (⅗ in).

Description: Color variable from white to yellow-gray with irregular reddish-brown maculations. Shell shape fusiform. Sculpture of teleoconch with axial cords most pronounced on the second-to-last whorl and faded toward the base. Spiral sculpture of spiral threads that become larger toward base. First whorl on protoconch smooth and round; remaining protoconch whorls with numerous axial riblets. Axial ribs on teleoconch whorls more separated and larger. Aperture narrow with toothlike structures on inner portion of outer lip. Columella with faded obsolete teeth and narrow parietal shield that resembles a keel. **Habitat:** Grass beds and sandy bottoms at depths from 0 to 46 m (150 ft). **Remarks:** Do not confuse with *C. floridana,* whose coloration may be similar, but sculpture is different. **Synonym:** *Anachis semiplicata* Stearns, 1873.

Costoanachis translirata (Ravenel, 1861)   **Well-ribbed Dovesnail**

Distribution: Maine to Florida, Texas; Yucatán.

Size: 23 mm (1 in).

Description: Color variable from different markings of yellowish to chestnut brown with a whitish band below suture and middle of body whorl. Shell shape elongate-conic, with flat-sided convex whorls. Sculpture of raised axial cords, crossed by weaker spiral ribs, giving a beaded appearance. Spiral threads strongest at base of body whorl. Protoconch smooth. Spire tall, more than half the length of shell. Aperture angled posteriorly, somewhat narrow but wider than most _Costoanachis_ spp. Outer apertural lip thickened with distinct dentition on inner surface. **Habitat:** Shell found in beach drift. Living snails found offshore. Shells found from shoreline to a depth of 152 m (500 ft). **Remarks:** Shells are more common on beaches from the upper coast of Texas. **Synonym:** _Anachis lafresnayi_ Fischer and Bernardi, 1857.

Parvanachis obesa (C. B. Adams, 1845)   **Fat Dovesnail**

Distribution: Virginia to Florida to Texas; Uruguay; Bermuda.

Size: 5 to 6 mm (1/ to ¼ in).

Description: Color variable from dull gray to yellowish-brown with irregular, spiral, flamelike bands of darker shades of brown, and sometimes a chocolate-colored spiral band at base. Shell shape fusiform, ovately obese. Sculpture of slightly convex whorls with ventricose axial ribs that terminate toward base. Spiral sculpture of riblets found between axial ribs, but entire at basal area. Aperture narrowly oblique and denticulate on inner surface of thickened outer lip. **Habitat:** In beach drift; bays to inlets in oyster shells or under hard substrates. **Remarks:** Often confused with _P. ostreicola,_ which differs by color pattern and stronger spiral sculpture.

Fasciolariidae

Fasciolariidae is a large family of gastropods comprising some 200 species that typically inhabit tropical to temperate waters. Fasciolariids are characterized by their medium- to large-sized (13 to 610 mm, or ½ to 24 in), fusiform, high-spired, shouldered shells with a well-developed, relatively long siphonal canal. Fasciolariids are carnivores; larger ones feed on bivalves and gastropods, whereas smaller species feed primarily on polychaetes. However, cannibalism has been reported in some species. In Texas Fasciolariidae is represented by 4 genera and 8 species whose size ranges from 76 to 610 mm (3 to 24 in); 2 species are discussed here.

Fasciolaria lilium Fischer, 1807   **Banded Tulip**

Distribution: North Carolina to Yucatán; Texas.
Size: 103 mm (4 in).

Description: Color yellowish-gray with irregular axial splotches of orangish-brown and spiral, purplish-brown lines. Shell shape spindlelike. Sculpture of teleoconch smoothish with convex whorls. First whorl on protoconch smooth and bulbous, with axial ribs on remaining protoconch whorls. Aperture ovate, parietal shield narrow and smooth, and siphonal canal relatively long. **Habitat:** Sand or muddy sand bottoms at depths from 4 to 46 m (13 to 150 ft).
Remarks: Subspecies are determined by sculpture, color, and shape.

Triplofusus giganteus (Kiener, 1840)   **Horse Conch**

Distribution: North Carolina to Florida, Texas to Yucatán. **Size:** 610 mm (24 in).

Description: Color yellowish-brown with a brown periostracum. Shell shape spindlelike. Sculpture of numerous wavy spiral cords, crossed axial ridges, and fine, irregular axial growth lines; teleoconch whorls distinctly shouldered. Aperture broadly ovate, and outer lip relatively thin and overlaps columella. Siphonal canal long and narrow. **Habitat:** On sand or muddy sand at depths from low-tide line to 200 m (656 ft). **Remarks:** The largest gastropod in the northwest Atlantic, including Gulf of Mexico. **Synonym:** *Pleuroploca gigantea* (Kiener, 1840).

Melongenidae

Melongenidae is a relatively small family of gastropods of approximately 25 species found in temperate and tropical marine waters. Melongenids are characterized by having moderate- to large-sized (25 to 584 mm, or 1 to 23 in), pear-shaped to torpedo-shaped shells with a broad body whorl, and various types of sculpture. Melongenids are one of the oldest neogastropod families and date from the Lower Cretaceous. Melongenids are typically found in nearshore sand and mud bottoms. They are usually regarded as scavengers, feeding on carrion; however, some are known to prey exclusively on bivalves. In areas where melongenids are abundant, several are often seen feeding on larger prey. In Texas Melongenidae is represented by 2 genera and 4 species whose size ranges from 100 to 232 mm (4 to 9 in); the 2 most common species are discussed here.

Distribution: Louisiana to Texas to Mexico.
Size: 200 mm (8 in).

Description: Color light brown to yellowish with dark brown axial bands that correspond with the shoulder spines of the body whorl, and with a broad, spiral, white band on the periphery of the body whorl. Shell left-handed and pear shaped. Sculpture of spiral threads. Spire turreted and only ⅕ the height of the shell. Aperture pear shaped, and siphonal canal long. **Habitat:** Infaunal. Just offshore in bays. **Remarks:** Although this shell is well known, much controversy surrounds the correct scientific name for the lightning whelk. For further clarification, please refer to Tunnell et al. (2010: 223). **Views:** 1) apertural view; 1a) top view; 1b) dorsal view. **Synonyms:** Mistakenly called *B. perversum* Linnaeus, 1758; *B. perversum pulleyi* Hollister, 1958.

Busycotypus spiratus (Lamarck, 1816)   **Pearwhelk**

Distribution: Texas to Yucatán.

Size: 100 mm (4 in).

Description: Color whitish-yellow with irregular axial, wavy, narrow bands of reddish-brown. Shell pear shaped and right-handed. Spiral sculpture of wavy threads. Axial sculpture of short, fine riblets found between spiral threads that are most prominent on the upper whorls. Beaded shoulders most distinct on the whorls following the protoconch and almost vanish on the body whorl. Spire short, tapered, and steplike. Protoconch smooth and round. Aperture pearlike. Upper portion of the columella with spiral threads, and lower region smooth. **Habitat:** Intertidal on sand to a depth of about 46 m (150 ft). **Remarks:** Several subspecies have been erected (named) because of variation in sculpture, spire, and strength of ridges.

Nassariidae

Nassariidae is a large family of gastropods comprising approximately 350 species. They are found in all tropical and temperate marine waters. Nassariids are characterized by their relatively small size and high-spired, fusiform, shouldered shell. The shell is usually axially and/or spirally sculptured or smooth. Axial sculpture is made up of broad to narrow ribs that are most pronounced posteriorly on the whorls. Spiral sculpture may be of smaller ribs or cords of comparable size to the axial cords, producing a cancellate appearance. Nassariids are usually found in shallow, marine, soft bottoms. Most are carrion feeders; however, some may become herbivores under certain conditions. In Texas Nassariidae is represented by 1 genus and 4 species whose size ranges from 6 to 12 mm (¼ to ½ in); 2 species are discussed here.

Nassarius acutus (Say, 1822) — Sharp Nassa

Distribution: Florida to Texas.
Size: 15 mm (⅗ in).

Description: Color dull yellowish-white with suffusions of darker orangish-brown and broken, brown narrow bands below the suture; protoconch orangish-brown. Shell shape elongate-conic. Sculpture of axial and spiral cords that produce elongate beaded nodules of varying size where they intersect, with most pronounced nodules above body whorl. Protoconch round and smooth. Aperture ovate with an anal notch and short siphonal canal. **Habitat:** Sandy bottoms from below tide line to a depth of about 110 m (360 ft). **Remarks:** Known as a carrion feeder but also feeds on molluscan egg cases.

Nassarius vibex (Say, 1822) — Bruised Nassa

Distribution: Massachusetts to Florida, Texas to Yucatán to Brazil.
Size: 20 mm (⅘ in).

Description: Color whitish-gray with spiral brown bands over teleoconch whorls. Shell shape broadly ovate. Sculpture of strong axial cords crossed by spiral ribs, giving a nodulose appearance; finer spiral threads toward the suture. Spire extended. Aperture ovate, with a thickened outer lip smooth above, crenulate toward short siphonal canal. Parietal ridge somewhat flared and smooth, with slight anal notch at the top and inner portion of the aperture. **Habitat:** Bays and sandy areas near low-tide line. Depth range 0 to 31 m (102 ft). **Remarks:** Scavenger. In Texas commonly found on tidal flats of bays.

Volutidae

Volutidae is a large family of gastropods made up of approximately 200 species. They are found in all seas, but the majority are generally located in tropical waters. Size of shells varies from small to large (9 to 500 mm, or ⅖ to 20 in). Shell may be ovately globose to narrowly fusiform. Patterning on the shell surface may be elaborate, highly colored and coated with a polished glazed surface, unicolored, or banded. Protoconchs usually consist of 1 to 3 whorls. The aperture is elongate and usually more than half the length of the shell. The siphonal canal is well developed. Anatomical characters such as the mantle, foot, and proboscis are highly colored and sometimes match the color of the shell. Feeding is believed to be carnivorous on other mollusks. In Texas, Volutidae is represented by 1 genus and 5 species whose size ranges from 126 to 197 mm (5 to 8 in); 1 species is discussed here.

Scaphella junonia (Lamarck, 1804) Junonia

Distribution: North Carolina to Florida to Texas; Mexico.
Size: 126 mm (5 in).

Description: Color ivory to cream with irregularly sized and squarish-shaped reddish-brown spots, with inner region of outer lip yellowish. Shell shape elongate-oval. Sculpture of slightly convex whorls. Teleoconch appears smoothish with fine, almost obscure spiral and axial growth threads. Protoconch dome shaped with first whorl smooth and brownish and following whorls with axial riblets and fine spiral threads that diminish on teleoconch whorls. Base of shell spirally ribbed. Aperture long, almost ¾ length of shell and pointed on both ends. Outer lip somewhat thin, and columella with ascending spiral ridges, most pronounced toward base. **Habitat:** Sand bottoms at depths from 0 to 110 m (360 ft). **Remarks:** The name of this elaborately colored shell comes from the peacock, or the "bird of Juno." It is rare in Texas.

Olivellidae

Olivellidae is a family of small-sized mollusks. The family was classified as Olivellinae, one of the 4 subfamilies of Olividae; however, because of significant differences the subfamily was elevated to family level. The shells resemble the true olives in shape and high gloss but differ significantly in size (smaller) and length of the aperture, which is only ⅓ to ½ the total length of the shell; the aperture in the true olives is about ¾ the total length of the shell. Olivellids are most commonly found in tropical and subtropical regions, with common species extending their range into temperate realms. They are located intertidally, in shallow subtidal areas, in sand or mud. They are carnivores on small prey, including bivalves, and some are scavengers. In Texas, Olivellidae is represented by 1 genus and 5 species whose size ranges from 11 to 17 mm (½ to ¾ in); 2 species are discussed here.

Olivella dealbata (Reeve, 1850)

Whitened Dwarf Olive

Distribution: Florida to Texas to the Caribbean.
Size: 14 mm (½ in).

Description: Color variable, whitish with reddish-brown axial flammules; above the white basal band are bands or arrow-shaped splotches on body whorl; incised suture distinctly reddish-brown. Shell shape elongate-oval. Sculpture smooth. Body whorl slightly convex. Spire extended and pointed at apex. Protoconch smooth and round. Aperture triangularly elongate, and columella has toothlike structures from base of aperture to the bottom of distinct axial color bands. **Habitat:** Mudflats and sandy areas in inlets of bays and beaches. Found at depths from low-tide line to 15 m (50 ft). **Remarks:** Coloration variable, but the apex and base are typically whitish except for the sutural area.

Olivella minuta (Link, 1807)   **Minute Dwarf Olive**

Distribution: Texas; Caribbean to Brazil.

Size: 15 mm (⅗ in).

Description: Color variable, basically dull white with purplish zigzag lines and/or bands sometimes suffused with orange or yellow; suture brownish-purple and grooved. Shell shape elongate-oval. Sculpture smooth. Body whorl somewhat convex. Spire elevated and pointed at apex. Aperture triangular; upper outer lip has a channel where lip meets body whorl. Columella with weak diagonal toothlike structures, parietal area smooth, and siphonal notch distinct. Bottom of color band obvious. **Habitat:** Infaunal. Sandy areas of surf zone to a depth of 15 m (50 ft). **Remarks:** _O. minuta_ is a nocturnal carnivore found burrowing just below the surface of the sand. **Synonym:** _Oliva verreauxii_ Duclos, 1857.

Olividae

Olividae is a large family of gastropods comprising some 300 species. They are typically located in shallow tropical seas around the world with some in temperate waters and a few in deep water and cold water. In living olivids the mantle and foot extensions cover the shell, thus providing shell collectors with smooth, polished, variously colored shells. The olivids have a high gloss and are usually brightly colored with a variety of color patterns. Most members of the family Olividae are carnivores or scavengers. In Texas, Olividae is represented by 1 genus and 2 species whose size ranges from 47 to 91 mm (2 to 4 in); only the most common species is discussed here.

Oliva sayana Ravenel, 1834 **Lettered Olive**

Distribution: North Carolina to Florida to Texas, Gulf of Mexico; Brazil.

Size: 91 mm (4 in).

Description: Color yellowish-gray with irregular zigzag markings of purplish-brown, with 2 broad spiral bands of darker coloration on the body whorl. Shell shape elongate-cylindrical. Sculpture smooth. Body whorl about 9/10 shell length. Spire smooth and straight sided. Aperture long and narrow, and with a sharp outer lip. Siphonal notch distinct, and columella with a few, strong, diagonal teeth. **Habitat:** In sand, in inlets, and offshore to depths of about 46 m (150 ft). Depth range 0 to 130 m (427 ft). **Remarks:** Live animals are seen burrowing just below the surface of the sand. **Synonym:** *O. sayana texana* Petuch and Sargent, 1986.

Marginellidae

Marginellidae is a large family of gastropods with over 400 species found in tropical and temperate climates. Marginellids vary in size and can be tiny to large. They are characterized by a short to flat spire, a large body whorl, and smooth shell surface. They can be broadly ovate, subcircular, to oblong. The shell is usually without sculpture and glossy. Sculpture, when present, is typically restricted to the shoulders in the form of axial striations. The aperture is generally narrow to flaring with several strong plicae (folds) on the columella. Marginellid habitats vary and range from the lower intertidal zone to deep water (1000 m, or 3300 ft) and from rocky reefs to soft shores. Feeding habits are not well known, but in all probability marginellids are carnivorous and feed on a variety of animal food. In Texas, Marginellidae is represented by 3 genera and 4 species whose size ranges from 8 to 14 mm (⅓ to ½ in); 1 species is discussed here.

Prunum apicinum (Menke, 1828)

Common Atlantic Marginella

Distribution: North Carolina to Florida to Texas, Gulf of Mexico; Caribbean.

Size: 14 mm (½ in).

Description: Color opaque white. Shell shape ovate. Sculpture of shell smooth. Spire slightly elevated and pointed. Aperture with a slight anal notch at the top, followed by a narrow channel that widens toward the base. Outer lip thick and smooth. Lower portion of columella usually with 4 distinct extended toothlike structures. **Habitat:** Found in bays south of Matagorda Island; in sand, clay, and mud environments; associated with seagrass. Found from shoreline to a depth of about 25 m (82 ft). **Remarks:** This species is rarely found alive in Texas. Abbott (1974) states approximately 1 in every 5000 specimens is left-handed. **Synonym:** _Marginella apicina_ Menke, 1828.

Cancellariidae

Cancellariidae is a somewhat large family of gastropods made up of about 150 species. Most inhabit tropical to subtropical climates, but some reside in cold waters. They are distinguished by their small to medium-sized, ovate to fusiform shells, with axial and spiral sculpture that produces a cancellate or reticulate surface, hence the family name. Feeding habits of the cancellariids are not well known; some authors hypothesize they feed on soft-bodied microorganisms, and others believe the internal anatomy is designed for suctorial (having structures for sucking) feeding. In Texas, Cancellariidae is represented by 2 genera and 2 species whose size ranges from 37 to 55 mm (1½ to 2¹/₁₂ in); 1 species is discussed here.

__Cancellaria reticulata__ (Linnaeus, 1767) **Common Nutmeg**

Distribution: North Carolina to Florida, Texas; Caribbean to Brazil.

Size: 55 mm (2$\frac{1}{12}$ in).

Description: Color variable; can have a base color of white to brown with axial or spiral splotches and/or bands of different shades of brown; protoconch yellowish-brown, smooth, and bulbous. Shell shape broadly ovate. Sculpture of strong axial cords and weaker spiral ribs that produce a cancellate, beaded appearance where they intersect. Whorls that follow the protoconch up to the body whorl have 4 spiral beaded cords each. Body whorl may have 12 to 16 beaded cords. Aperture semilunar; outer lip crenulate and interior denticulate. Columella with large toothlike structures and a distinct parietal shield. **Habitat:** In sand and grass beds from shoreline to a depth of 73 m (240 ft). **Remarks:** In Texas this snail is common offshore.

Terebridae

Terebridae is a family of gastropods known as auger shells that comprises 6 genera and approximately 150 species. Terebrids are found in all tropical and subtropical waters of the world. Terebrid shell form is the most consistent of the 3 conoidean families; it is also the most stenotypic, geographically and ecologically, of these families. It occurs primarily on sand in shallow, subtidal depths. The shells are tall and slender, conic, and many whorled, with a pointed spire and relatively small body whorl. The small aperture has a short siphonal canal with a short but distinct anal sinus above and a basal notch below. Most terebrids are infaunal and will burrow to depths usually not exceeding their length. However, some species occur just below the surf zone on sandy beaches. The terebrid shell is thick and quite effective against predation, especially from the Calappidae family of crabs. Many terebrids are host specific and feed on only 1 species of polychaete. In Texas, Terebridae is represented by 2 genera and 7 species whose size ranges from 18 to 152 mm (¾ to 6 in); 5 species are discussed here.

Hastula maryleeae Burch, 1965 — Marylee's Auger

Distribution: Texas to British Honduras.
Size: 25 mm (1 in).

Description: Color variable, typically dark brown with a white band and obscure white markings on body whorl. Shell shape tall, slender, and tapers to a point. Sculpture of thin axial ribs that begin at the suture of each whorl and end about ⅓ the way down of each whorl. Apex of 2 glossy pointed whorls. Aperture ovate, and outer lip thin and sharp. Siphonal notch recurved, with a pointed structure at the base of siphonal canal. **Habitat:** Sandy surf zone. **Remarks:** Named after Marylee Burch, who collected the holotype at Surfside Beach, Freeport, Texas. **Synonyms:** *Terebra maryleeae* Burch, 1965; *T. tobagoensis* Usticke, 1969.

Hastula salleana (Deshayes, 1859) — Salle's Auger

Distribution: Florida, Texas; Veracruz, Mexico; Brazil.
Size: 25 to 40 mm (1 to 1½ in).

Description: Color variable, grayish-brown to pinkish-gray with a white band just below suture, a larger interrupted brown band below the white band, and a slender whitish band on base of body whorl. Shell shape tall, slender, and tapers to a point. Sculpture of axial ribs that begin at suture and end almost halfway down each whorl. Spire pointed. Aperture shape ovate, with a thin outer lip. Siphonal notch recurved and pointed. **Habitat:** Sandy surf zone. **Remarks:** *H. salleana* is usually found associated with *H. maryleeae*. When this snail is interrupted by wave action, it quickly buries itself in the sand.

Terebra dislocata (Say, 1822)   Common American Auger

Distribution: Maryland, Florida, Texas; Caribbean; Brazil.
Size: 38 to 64 mm (1½ to 2½ in).

Description: Color variable bluish to yellowish. Shell shape elongate-conic. Sculpture of axial ribs found just below the shoulder. Axial sculpture of intermittent long ribs. Spiral sculpture of threads found between axial ribs. Protoconch smooth and pointed. Aperture oblong, and siphonal canal short and twisted to the left. Outer area of the columella with a raised ridge.
Habitat: Sandy bottoms at depths from shoreline to 146 m (480 ft).
Remarks: Can be found in clumps of sand at low tide.

Terebra protexta (Conrad, 1846)   Fine-ribbed Auger

Distribution: North Carolina, Florida, Texas; Brazil.
Size: 20 to 28 mm (⅘ to 1⅛ in).

Description: Color brownish. Shell shape elongate-conic. Sculpture of bladed-type axial cords indented just below the suture. Sculpture of fine, evenly spaced spiral threads between axial cords. Suture well defined. Protoconch whorls smooth. Shape of aperture oval and with a thin outer lip. Columella calloused and with a ridge that twists to the left. **Habitat:** Sandy bottoms just offshore at depths from 1 to 106 m (3 to 350 ft).
Remarks: This auger has been collected alive at San Luis Pass.

__Terebra taurina__ (Lightfoot, 1786)   **Flame Auger**

Distribution: Florida, Texas; Caribbean; Brazil.

Size: 102 to 178 mm (4 to 7 in).

Description: Color yellowish-white with broad reddish-brown flammules. Shell shape large, tall, and tapers to a point. Sculpture of heavily knobbed teleoconch whorls with the middle whorls less knobbed. C-shaped axial threads on body and penultimate whorls. Shape of aperture semilunar. Outer lip thin, convex, and somewhat sharp. Siphonal canal wide, anal canal pointed, and columella turned backward. **Habitat:** Sandy bottoms offshore. Found alive offshore at depths from 1 to 80 m (3 to 260 ft). **Remarks:** This colorful shell is usually collected onshore after dredgings. **Synonyms:** *T. flammea* Lamarck, 1822; *T. texana* Dall, 1898.

Turridae

Turridae, the largest family of mollusks, includes more than 600 genera and more than 11,000 described species worldwide, including many fossil species, and many more waiting to be discovered and described. Recently, the broad family Turridae has been divided into 13 separate families, based on molecular and morphological characters, but in this field guide we follow the classification used in Tunnell et al. (2010) and keep all "turrids" in the single family Turridae. In Texas Turridae is represented by 30 genera and 54 species whose size ranges from 5 to 101 mm (⅕ to 4 in), but only the 4 most common species are discussed here.

Cryoturris cerinella (Dall, 1889)

Little Waxy Mangelia

Distribution: South Carolina, Florida, Texas.
Size: 8 to 12 mm (⅓ to ½ in).

Description: Color golden, yellowish to white. Shell shape elongate, towerlike. Sculpture of somewhat flowing axial cords. Spiral sculpture on base distinct. Protoconch of smooth, rounded whorls. Aperture U-shaped, siphonal canal short, and the turrid notch readily seen. **Habitat:** Found alive on Stetson Bank and dead on Texas beaches. Depth range 0 to 73 m (240 ft). **Remarks:** Identified by its towerlike appearance. **Synonym:** *Kurtziella cerinella* (Dall, 1889).

Kurtziella cerina (Kurtz and Stimpson, 1851)

Waxen Mangelia

Distribution: Massachusetts, Florida, Texas; Yucatán.
Size: 10 mm (⅖ in).

Description: Color tannish-white. Shell shape elongate, towerlike. Sculpture of axial cords rounded at the shoulders, producing a wavy appearance. Spiral sculpture of numerous spiral threads, pronounced toward the base. Narrow channel with distinct spiral lines between body whorl and preceding whorl. First whorl of protoconch smooth, and last protoconch whorl with axial riblets. Aperture wide, and siphonal canal and turrid notch short. **Habitat:** Sandy and calcareous environments at depths from 1 to 55 m (3 to 180 ft). **Remarks:** Closely related to *K. atrostyla*. **Synonym:** *Mangelia cerina* (Kurtz and Stimpson, 1851).

Polystira albida (Perry, 1811)   **White Giant Turrid**

Distribution: Florida, Texas; Caribbean.

Size: 76 to 116 mm (3 to 4½ in).

Description: Color white. Shell shape spindlelike. Sculpture of strong spiral cords and fine axial fringelike threads that angle right above and left below larger cords on older whorls; newer whorls with more cords. Body whorl with doubled primary cord that begins at turrid notch. Spire tall and pointed. Aperture shape oblong, and siphonal canal long and narrow. **Habitat:** Found on sandy bottoms at depths from 9 to 91 m (30 to 300 ft). Depth range 9 to 274 m (30 to 900 ft). **Remarks:** Dead shells found on beach typically come from dredged spoil material.

Pyrgocythara plicosa (C. B. Adams, 1850)   **Plicate Mangelia**

Distribution: Massachusetts to Texas.

Size: 6 to 13 mm (¼ to ½ in).

Description: Color tannish-brown to reddish-brown. Shell shape ovately towerlike. Sculpture of broad, rounded axial cords crossed by thinner spiral ridges, producing a cancellate appearance. Protoconch with rounded, smooth whorls. Aperture long and U-shaped, upper portion of lip thickened with distinct U-shaped turrid notch, and siphonal canal short. **Habitat:** In bays, found in mud and grass at depths from shoreline to 22 m (73 ft); alive from 0.5 to 4.0 m (2 to 12 ft). **Remarks:** Commonly found in beach drift in all Texas bays. This snail lives on old oyster shells, where specimens can be collected alive. **Synonym:** *Mangelia plicosa* C. B. Adams, 1850.

Architectonicidae

Architectonicidae is a family of gastropods comprising approximately 6 genera and a number of species found in warm waters of the world. They are found in shallow to deep water. Shells vary in shape from wide and conic to low and disklike. Shells are low spired and discoidal and may have smooth or ornamented shells. Sculpture may be of spiral grooves, ridges, and/or granulations. Architectonicids are difficult to study in the field because of their subtidal habitat; also they burrow into the sand and come out only at night to feed or spawn. Some architectonicids live under rocks, and others live among soft corals. Collectors value their unusual shells. The wide distribution of architectonicids is usually attributed to the extended planktonic larval stage of this family. In Texas, Architectonicidae is represented by 3 genera and 3 species whose size ranges from 3 to 51 mm (⅛ to 2 in); 1 species is discussed here.

Architectonica nobilis Röding, 1798   **Common Sundial**

Distribution: Florida, Texas.
Size: 51 mm (2 in).

Description: Color tannish-yellow with spiral rows of reddish-brown spots. Shell shape wide and conic. Sculpture of spiral cords and axial riblets that give a beaded appearance on upper but not on lower whorls. Spire short and smooth. Aperture subcircular. Base or bottom of shell flattened, with interrupted spiral cords. Umbilicus deep, with toothlike structures. **Habitat:** In sandy, shallow water of nearshore Gulf. Depth range 0 to 164 m (540 ft). **Remarks:** Shells common on Texas beaches, but they are popular among shell collectors.

Omalogyridae

The family Omalogyridae represents some of the smallest gastropods in the world and is placed in its own superfamily, Omalogyroidea. Information on these gastropods is limited, with most faunal studies excluding the less than 1 mm (⅒₅ in) size range for adult shells. The minute shells range in size from less than 2 mm to less than 0.5 mm (1/12 in to 500 µm). Shell coloration varies from translucent white to translucent reddish-brown, and most shells are planispiral. Sculpture may be smooth but is usually well ornamented with crenulated axial ridges. Omalogyrids are distributed around the world and are usually found in the lower intertidal and sublittoral zones of coral reefs. On the Texas coast Omalogyridae is represented by 6 species and 2 genera whose size ranges from 0.35 to 0.75 mm (1/65 to 1/38 in, or 350 to 750 µm); 1 species is discussed here.

Ammonicera minortalis Rolán, 1992

Minuscule Ammonicera

Distribution: Texas; Cuba; Canary Islands.

Size: 0.35 mm (1/65 in, or 350 µm).

Description: Color translucent reddish-brown. Shell shape planispiral and right-handed. Sculpture of numerous smooth axial ribs, with a row of spiral nodules produced by the axial ribs on both sides of shell. Aperture subcircular.

Habitat: Calcareous environments. Previously recorded depth ranges are 1 to 20 m (3 to 66 ft). **Remarks:** This is probably the smallest gastropod in Texas and is thought to be the smallest gastropod in the western Atlantic. Specimens have been collected at Flower Garden Banks and Alacrán Reef, Mexico, at depths from 4 to 46 m (12 to 154 ft) and are housed in the Tunnell collection. Specimen in photograph is from Flower Garden Banks.

Murchisonellidae

Murchisonellidae is a small family of tiny gastropods with only 2 genera and 3 species listed for the Gulf of Mexico. Shell coloration is generally translucent white in fresh material to opaque white in older specimens. The protoconch is made up of a single, smooth, glasslike whorl that is tilted and somewhat immersed in the preceding whorl. The family is not well known. In Texas, Murchisonellidae is represented by 2 species; the most common is discussed here.

Henrya goldmani Bartsch, 1947

Distribution: Florida, Texas.
Size: 2.2 mm (¹⁄₁₂ in).

Description: Color of shell translucent white. Shell shaped like a cocoon. Sculpture smoothish. Whorls rounded; suture distinct. Protoconch round, smooth, and immersed into the succeeding whorl. Aperture ovate; outer lip thin and flared at base.

Habitat: Hypersaline lagoons to coral reefs. Previously recorded depth range 0 to 4 m (13 ft). **Remarks:** Sometimes mistaken for *H. morrisoni* Bartsch, 1947. Specimen in photograph collected from Flower Garden Banks at a depth of 23 m (75 ft).

Pyramidellidae

Pyramidellidae is a large family of gastropods made up of as many as 6000 species found throughout the world. The taxonomy of this family is in flux. These small gastropods are found from the intertidal zone to several thousand meters. Live animals are rare, but shells are numerous and ubiquitous. Identification of pyramidellids has relied primarily on shell characters such as ornamentation, columellar folds, and the protoconch. Changes in life stages and lack of anatomical characters have probably led to numerous errors in taxonomy. Other problems associated with identification are the small size and variation in shell morphology within species. Authors do not always agree on decisions that constitute a genus, subgenus, and species. Even placement of this family within subclasses has been controversial. Shells range in size from less than 3 to about 54 mm (⅛ to 2¹⁄₁₀ in), with a few larger pyramidellids. The shells are rarely planispiral; most are trochiform, ovate to elongate. The pyramidellids are ectoparasites that feed on body fluids of a variety of invertebrate species but primarily polychaete worms and other mollusks. Pyramidellids feed by attaching themselves to the host by using an oral sucker. Most pyramidellids stay near their host; most are not host specific, but some are. Pyramidellids are hermaphroditic, possessing sperm in the gonads as well as ripe ova within the same tubules. Most pyramidellids do not have a common name. On the Texas coast Pyramidellidae is represented by 18 genera and 58 species whose average size ranges from 3 to 14 mm (⅛ to ½ in); the most common 17 species are discussed here.

Boonea impressa (Say, 1822) **Impressed Odostome**

Distribution: New Jersey, Massachusetts, Florida, Texas; Mexico; Brazil.

Size: 7 mm (³⁄₁₀ in).

Description: Color of shell opaque. Shell shape conic. Sculpture of strong spiral ribs. Whorls separated by distinct suture. Protoconch round, smooth, and partially immersed into succeeding whorl, which produces a half-beadlike appearance. Aperture ovate; outer lip flared at base, with distinct columellar tooth or fold found on older specimens. **Remarks:** This species is parasitic on a variety of hosts, especially _Crassostrea virginica_. _B. impressa_ is typically found in bays on oyster reefs. Those found offshore on reefs and banks are indicative of Pleistocene-Holocene drowned coastal bays. Genus named for Texas conchologist Connie Boone. _B. impressa_ displays 2 known forms, as seen in figures 1 and 2.

Boonea seminuda (C. B. Adams, 1839) **Half-smooth Odostome**

Distribution: New Jersey, Massachusetts, Florida, Texas; Mexico; Brazil.
Size: 7 mm (³⁄₁₀ in).

Description: Color off white. Shell shape tall and conic. Sculpture of spiral and axial ribs that gives a cancellate, beaded appearance where they intersect. Body whorl with upper half cancellate, lower half with spiral cords, and crowded axial threads in inner spaces. Protoconch tilted, smooth, bulbous. Aperture ear shaped, and outer lip crenulate. **Habitat:** Ectoparasite on a variety of hosts, usually on the Atlantic bay scallop (_Argopecten irradians_). Depth range 0 to 101 m (330 ft). **Remarks:** Other hosts are common Atlantic slippersnail (_Crepidula fornicata_) and eastern white slippersnail (_C. depressa_).

Cyclostremella humilis Bush, 1897

Distribution: North Carolina, Florida, Texas; Mexico.
Size: 2 mm (¹⁄₁₂ in).

Description: Color translucent white. Shell shape planispiral. Sculpture of crowded, S-shaped, radial lines with microscopic spiral threads. Tilted sutures form a channel. Aperture not totally circular. Protoconch somewhat immersed into the succeeding whorl. Umbilicus wide and deep. **Habitat:** Probably a surf-zone inhabitant. Depth range 0 to 29 m (95 ft). **Remarks:** Species first placed in the family Vitrinellidae (now Tornidae), later moved to the family Cyclostremellidae, and finally placed in the family Pyramidellidae because of morphological and anatomical features, with emphasis on the features of the protoconch. This shell is commonly found in beach drift all along the Texas coast.

Eulimastoma canaliculatum (C. B. Adams, 1850)

Distribution: Florida, Texas; Mexico; Brazil.
Size: 3 mm (⅛ in).

Description: Color yellowish-white. Shell shape variable but typically straight sided, tall, and conic. Sculpture variable but usually smoothish with grooved sutures of varying depths, and may be keeled at periphery of body whorl. Development of toothlike structures on columella and development of umbilicus vary. Aperture subcircular in shape. **Habitat:** Variable, from bays to deep water to a depth of about 128 m (420 ft). **Remarks:** This species is quite variable morphologically; varies from long and slender to short and broad. **Synonym:** *Odostomia canaliculata* C. B. Adams, 1850.

Eulimastoma harbisonae Bartsch, 1955

Distribution: Texas.
Size: 2 mm (¹⁄₁₂ in).

Description: Color whitish. Shell shape ovoid-conic. Sculpture simple, smoothish with rounded whorls and deep sutures. Protoconch deeply immersed into succeeding whorl. Aperture ear shaped, and columella distinctly toothed. **Habitat:** Texas bays. Possibly the dominant gastropod in Copano Bay. Depth range 0 to 2 m (6 ft). **Remarks:** *E. harbisonae,* like most micromollusks, does not have a common name; it produces specimens with an open and a closed umbilicus, which have been differently named *E. harbisonae* and *E. olssoni,* respectively. **Synonym:** *E. olssoni* Bartsch, 1955.

Eulimastoma teres (Bush, 1885)

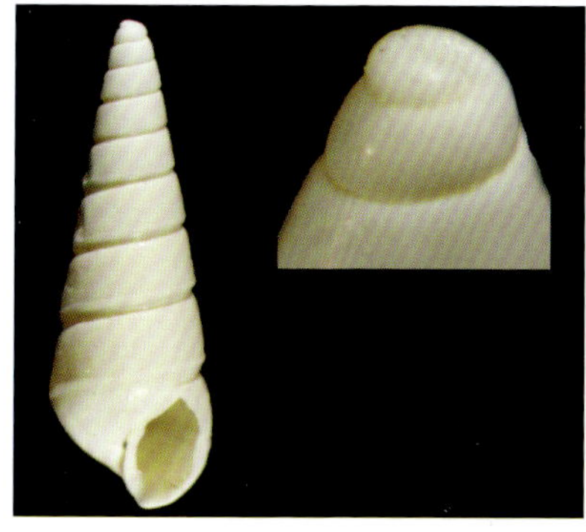

Distribution: North Carolina, South Carolina, Texas.
Size: 5 mm (⅕ in).

Description: Color yellowish-white. Shell shape variable, usually tall, slender, and tapers to a soft point. Sculpture simple, smoothish with channel-like sutures, and keeled at base of the lower whorls. Protoconch immersed into succeeding whorl. Aperture ear shaped to diamond shaped. Columella angled, with toothlike structures. **Habitat:** Texas bays. Depth range 0 to 360 m (1180 ft). **Remarks:** *E. teres* may be a synonym of *E. engonium* but lacks paired ridges at the sutures and columella teeth, and *E. engonium* is usually found offshore.

Evalea emeryi (Bartsch, 1955)

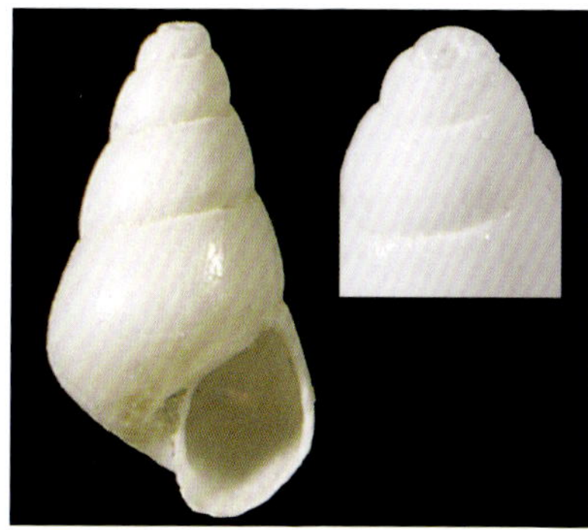

Distribution: North Carolina, Texas.
Size: 3 mm (⅛ in).

Description: Color yellowish-white. Shell shape somewhat elongate and conic. Sculpture simple, with rounded whorls, impressed sutures, and an inflated body. Protoconch immersed into the succeeding whorl. Aperture ear shaped, with thin outer lip that flares at the base. **Habitat:** Bay species. Shells found at depths from 0 to 58 m (190 ft). **Remarks:** Rarely shows a columellar tooth.

Fargoa bushiana (Bartsch, 1909)

Distribution: Massachusetts, North Carolina, Florida, Texas. **Size:** 4 mm (⅙ in).

Description: Color yellowish-white. Shell shape somewhat tall and conic. Sculpture typically of 3 spiral inflated cords crossed by axial cords, producing large beaded structures and somewhat squarish inner spaces. Has 1 smooth, raised, flattened spiral cord above the suture. Protoconch immersed into the succeeding whorl. Aperture ovate to pear shaped, and columella with toothlike structure created from the raised spiral cord. **Habitat:** Thought to parasitize tube worm polychaetes. Found at depths from 1 to 91 m (3 to 300 ft). **Remarks:** Shells common in beach drift.

Fargoa gibbosa (Bush, 1909)

Distribution: Massachusetts, North Carolina, Florida, Texas. **Size:** 5 mm (⅕ in).

Description: Color translucent whitish-gray. Shell shape globose with a strong tapering conical upper end. Sculpture simple, smooth with convex whorls and distinct sutures. Aperture ovate; outer lip thin, flares at base; and columella with distinct toothlike structure. **Habitat:** In bays, inlets, and shallow shelf areas of Texas coast. Found at depths from shoreline to 57 m (187 ft). **Remarks:** Found in beach drift or dredged material from the shallow shelf area of Texas coast.

Houbricka incisa (Bush, 1899)

Distribution: Georgia, Florida, Texas; Yucatán.

Size: 8 mm (⅓ in).

Description: Color orangish-white with a brownish band toward bottom of each whorl; protoconch translucent white. Shell shape elongate, slender, towerlike. Sculpture of convex whorls with many rounded axial ribs that cross short, thin, spiral grooves. Lower part of the body whorl with fine spiral threads. Protoconch smooth, bulbous, partially immersed with oldest part exposed as if lying on its side. Aperture ovate with base slightly flared. **Habitat:** Found in sand, mud, and algae. Occurs all along the Texas coast at depths from 0 to 55 m (180 ft). **Remarks:** This pyramidellid is quite tall and narrow, almost towerlike. **Synonym:** *Turbonilla constricta* Bush, 1899.

Longchaeus suturalis (Lea, 1843)

Distribution: North Carolina, Florida, Texas; Yucatán; Costa Rica.

Size: 14 mm (½ in).

Description: Color tannish. Shell shape tall, slender, and conic. Sculpture of somewhat smooth, flat-sided whorls with a narrow channel. Protoconch small, smooth, bulbous, and immersed into succeeding whorl. Body whorl with incised cord at the periphery. Aperture ear shaped; columella with 1 strong, rounded, toothlike structure and 2 ridges toward the slightly flared base. **Habitat:** Ectoparasite; in inlets and lagoons, sand, mud, and grass bottoms. Depth range 0 to 49 m (160 ft). **Remarks:** One of the largest pyramidellids on the Texas coast. **Synonyms:** *Pyramidella suturalis* Lea, 1843; *P. crenulata* Holmes, 1859.

Peristichia toreta Dall, 1889

Distribution: North Carolina, Florida, Texas; Campeche Bank, Mexico.
Size: 14 mm (½ in).

Description: Color yellowish-white. Shell shape tall, slender, towerlike. Sculpture of spirally beaded cords with 1 spiral cord just above the suture. First beaded spiral cord smallest, second beaded cord medium, and third cord largest, but preceded by a spiral channel intersected by axial riblets. Protoconch smooth, bulbous, partially immersed, and tilted to one side. Aperture subcircular, with slightly crenulate outer lip recurved at base. **Habitat:** Sandy bottoms at depths from shoreline to 101 m (330 ft). **Remarks:** Collected in beach drift from South Padre Island and Galveston.

Sayella hemphillii (Dall, 1884)

Distribution: Florida, Texas.
Size: 5 mm (⅕ in).

Description: Color variable, typically reddish-brown with whitish bands located below suture. Shell shape variable from short cocoon-like to elongate and cocoon shaped, with a broad to tall body whorl. Sculpture smooth, except for irregular axial growth lines. Whorls somewhat rounded with convex sutures. Protoconch immersed into the succeeding whorl. Aperture ovate to ear shaped. Columella slightly folded backward toward base with base of apertural lip slightly flared. **Habitat:** Bays and estuaries; also in beach drift. Depth range 0 to 2 m (6 ft). **Remarks:** Shell shape and color variable; can be short and stout (specimen 1) to elongate and slender (specimen 2). **Synonym:** *S. livida* Rehder, 1935.

Turbonilla hemphilli Bush, 1899   **Hemphill's Turbonilla**

Distribution: Florida, Louisiana, Texas; Yucatán; Bermuda.

Size: 13 mm (½ in).

Description: Color translucent white to bluish-gray. Shell shape tall, slender, and towerlike. Sculpture of rounded axial riblets on each whorl. Axial riblets on body whorl terminate at base, thus providing a smooth base. Whorls convex with deep sutures. Protoconch pointed, bulbous, tilted to one side, and slightly immersed into the succeeding whorl. Aperture square shaped, with thin outer lip.
Habitat: In bays and offshore at depths from 0 to 137 m (450 ft).
Remarks: Hemphill's turbonilla, when collected alive in bays, is a bluish-gray; when collected offshore, it is white.

Turbonilla interrupta (Totten, 1835) **Interrupted Turbonilla**

Distribution: Canada; Massachusetts, North Carolina, Florida, Texas; Brazil.
Size: 8 mm (⅓ in).

Description: Color brownish-red, with the protoconch a darker shade. Shell shape tall, slender, and towerlike. Sculpture of axial ribs and interrupted spiral grooves. Whorls somewhat flat sided, with body whorl having distinct spiral threads on the base. Protoconch bulbous, partially immersed, and tilted to one side. Aperture subovate with thin outer lip and slightly flexed columella. **Habitat:** Ectoparasite along Texas shores at depths from 0 to 196 m (643 ft). **Remarks:** Holotype lost or misplaced. Sculpture variable in this species; transverse ribs can number from 20 to more than 30.

Turbonilla levis (C. B. Adams, 1850)

Distribution: North Carolina, Florida, Louisiana, Texas; Colombia; Jamaica.
Size: 8 mm (⅓ in).

Description: Color white in fresh shells, opaque in old shells. Shell shape tall, slender, and conic. Sculpture of rounded axial ribs on each whorl, with axial ribs of body whorl terminating toward base, leaving base smooth. Whorls convex and in steplike progression, with deep distinct sutures. Protoconch smooth, bulbous, tilted to one side, and partially immersed into the succeeding whorl. Aperture ovate and somewhat flared backward at the base. **Habitat:** Sandy mud bottoms at depths from 13 to 196 m (43 to 643 ft). **Remarks:** Found on shale domes such as Stetson Bank and rocky bottoms such as Seven and One-half Fathom Reef. **Synonym:** *T. crenulata* (Menke, 1830).

Turbonilla portoricana Dall and Simpson, 1901

Distribution: Texas, Louisiana; Colombia; Venezuela; Brazil. **Size:** 5 mm (⅕ in).

Description: Color white with a narrow yellowish-brown band just above sutures of fresh shells. Shell shape tall, slender, and towerlike. Sculpture of convex whorls with distinct sutures. Axial sculpture of transverse ribs crossed by spiral threads, giving cancellate appearance. Axial ribs continue onto columella, and spiral threads continue onto base. Protoconch smooth, bulbous, translucent, tilted to one side, and partially immersed into the succeeding whorl. Aperture subovate, with thin outer lip and angled columella. **Habitat:** Bays and inlet areas to offshore waters at depths from 10 to 129 m (33 to 423 ft). **Remarks:** Common along Texas coast.

Acteonidae

Acteonidae is a small family of gastropods made up of approximately 7 genera and about 60 species. Shells are variable in shape, from wide to narrowly ovate or oblong. The spire in most acteonids is conspicuous with a tilted nuclear whorl. The shell is relatively strong and is totally equipped to contain the retracted animal. The aperture is narrow and ovate. Acteonids are specialized predators of polychaete worms, and they seem to breed during summer months. Species of this family live in a variety of habitats ranging from sandy, shallow intertidal areas to depths exceeding 3000 m (9900 ft). In Texas Acteonidae is represented by 2 genera and 2 species whose size ranges from 3 to 10 mm (⅛ to ⅖ in); 1 species is discussed here.

Japonacteon punctostriatus (C. B. Adams, 1840)   **Pitted Baby-bubble**

Distribution: Massachusetts, North Carolina, Florida, Texas; Cuba, Yucatán.

Size: 8 mm (½ in).

Description: Color grayish-brown with a white band below the suture. Shell shape globosely oval. Sculpture smoothish; spire whorls smooth and rounded with incised sutures. Body whorl large, rounded at shoulder, flat sided toward the periphery, with spiral rows of small holes about midway down. Protoconch with first whorl slightly immersed into the succeeding whorl, as seen under high magnification. Aperture ovately elongate, outer lip thickened, lower lip flared, and columella angled with a strong ridge. **Habitat:** Mud or sandy bottoms at depths from low-tide line to about 110 m (360 ft). **Remarks:** Similar to Rehder's baby-bubble (see Tunnell et al. 2010: 277) but less globose and shorter, thinner outer lip, and the columella with weaker toothlike structure. **Synonym:** *Acteon punctostriatus* C. B. Adams, 1840.

Bullidae

Bullidae is a small family of gastropods made up of only 1 genus, *Bulla,* and approximately 30 species found in temperate and tropical seas. Shell shape is broadly oblong to ovate. Most shells are smooth, but a few may have fine spiral striations. The shell is ovoid and robustly calcified, smooth, and inflated with a deep, sunken spire. The aperture is at least as long as the shell and rounded anteriorly. Bullids are found buried in sediment during the day; at night, they come out to feed on algae and sometimes on other mollusks. On the Texas coast Bullidae is represented by 1 species.

Bulla striata Bruguière, 1792   Striate Bubble

Distribution: Texas, Louisiana, Florida; Cuba; Yucatán, Tabasco, Mexico; Brazil.
Size: 44 mm (1¾ in).

Description: Color tannish with mottlings of darker shades of brown; interior and parietal shield whitish. Shell shape broadly ovate. Axial sculpture of fine, irregular growth lines. Spiral sculpture at base and within sunken spire; sunken spire with a distinct perforation. Aperture long, slightly flared, and narrow toward the top and broadly flared and wide toward the base. Parietal shield narrow, and columella turned backward. **Habitat:** Seagrass beds in shallow water. Depth range 0 to 30 m (100 ft). **Remarks:** This species is a carnivore. Mantle covers shell when animal is active.

Haminoeidae

Haminoeidae is a relatively small family of gastropods of 12 genera and approximately 90 species found in temperate and tropical seas. Shells are thin and fragile, narrowly cylindrical to broadly ovate, translucent to almost transparent when alive, but usually opaque when dead. Coloration of haminoeids is variable and is thought to be determined by the type of environment each inhabits. For instance, those living in mud have a somber coloration, and those living in sand are lighter to translucent. Some species of haminoeids are able to change color to fit the substrate. The spire is sunken and may have a small perforation or be completely covered. The posterior, narrower part of the aperture usually extends beyond the spire. The outer lip of the aperture is typically longer than the shell itself. These animals are herbivores that feed on various types of algae. On the Texas coast Haminoeidae is represented by 2 genera and 5 species whose size ranges from 8 to 12 mm (⅓ to ½ in); 2 species are discussed here.

Haminoea antillarum (d'Orbigny, 1841)   **Antillean Paperbubble**

Distribution: Florida, Texas; Cuba; Yucatán; Brazil.
Size: 18 mm (¾ in).

Description: Color yellowish-white. Shell shape globose. Sculpture smoothish except for fine, irregular axial growth lines. Protoconch sunken, as if engulfed by body whorl. Aperture long, flared above, narrow in the middle, and broadly flared below. Lip of columella concave. **Habitat:** Epifaunal in inlet-influenced areas. Depth range 0 to 2 m (6 ft). **Remarks:** Typically collected on the beach. It is omnivorous and unable to totally withdraw into its shell. **Synonym:** *Bulla cerina* Menke, 1853.

Haminoea succinea (Conrad, 1846)   **Conrad's Paperbubble**

Distribution: Florida, Louisiana, Texas; Yucatán; Bermuda.
Size: 12 mm (½ in).

Description: Color white to pale yellow. Shell shape cylindrical. Sculpture of fine spiral and axial threads that give a frosted appearance. Body whorl almost straight sided. Protoconch sunken, as if engulfed by the body whorl. Aperture slightly flared toward the top and broadly flared toward the base. Lower lip slightly flared and recurved back, and columella angled and slightly turned back, producing a smoothish channel and chinklike umbilicus. **Habitat:** Collected alive on hard substrate covered with algae and in grassflats. Depth range 0 to 55 m (180 ft). **Remarks:** Shell fragile; not common in beach drift.

Cylichnidae

Cylichnidae (= Scaphandridae) is a relatively small family of gastropods comprising approximately 6 genera and 50 species found throughout the world. Cylichnids are usually found in colder waters or at great depths but also inhabit temperate and tropical waters at shallower depths. This family includes species that have retained many primitive characteristics. Although the number of specimens of cylichnids found dead in beach drift is high, knowledge of the anatomy of species within this family is weak. The shell may be cylindrical, ovoid, globose, pear shaped, or elongate. The spire may be depressed or sunken below the apex of the shell or project as a moderate spire. The outer surface of the shell is usually smooth or spirally striate, but in some species sculpture comprises punctate grooves. Cylichnids are found from intertidal depths to bathyal depths of 5500 m (18,150 ft). All species are infaunal and typically inhabit the upper centimeter of soft substrate. Cylichnids typically feed on sand-living invertebrates, worms, and other mollusks. In Texas, Cylichnidae is represented by 1 genus and 6 species whose size ranges from 3 to 6 mm (⅛ to ¼ in); 3 species are discussed here.

Acteocina canaliculata (Say, 1826)   **Channeled Barrel-bubble**

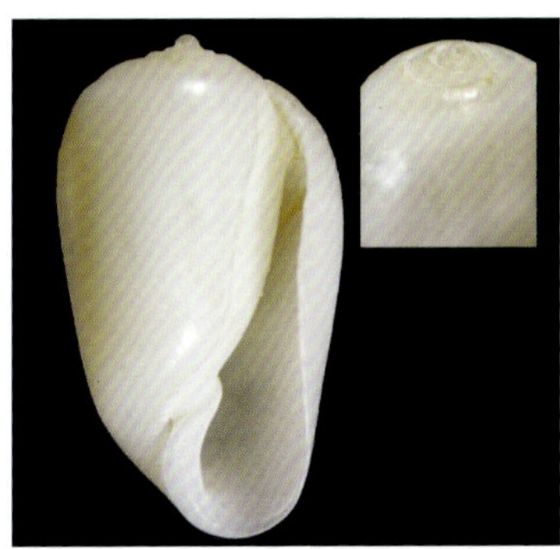

Distribution: Nova Scotia; Florida, Texas; Caribbean.
Size: 6 mm (¼ in).

Description: Color yellowish-white. Shell shape broadly cylindrical. Sculpture smooth except for fine, irregular axial growth lines; whorl preceding body whorl engulfed by body whorl, giving spire a sunken look; grooved spiral depression below suture with numerous axially curved riblets. Protoconch somewhat extended and curled or curved inward. Aperture long, narrow toward top, wide toward base, with a thin outer lip and lower part of lip slightly flared. Columella thickened and angled with a distinct, thickened spiral fold. **Habitat:** Coastal bays and inlets at depths from 0 to 49 m (160 ft). **Remarks:** Common all along the Texas coast. **Synonym:** *Retusa canaliculata* (Say, 1826).

Acteocina candei (d'Orbigny, 1841)

Distribution: North Carolina to Argentina; Texas.
Size: 5 mm (⅕ in).

Description: Color whitish. Shell shape elongate-cylindrical. Sculpture smooth; body whorl 9/10 length of shell; whorl preceding body whorl sunken into body whorl. Spire somewhat extended. Protoconch projects upward and curls down into the succeeding whorl. Aperture elongate, narrow toward top, wider toward base, with a thin outer lip, and columella with thick and twisted, channeled fold. **Habitat:** Sandy bottoms at depths from 0 to 51 m (167 ft). **Remarks:** Similar to the channeled barrel-bubble, the bay species. However, *A. candei* (no common name) is more cylindrically uniform, protoconch is larger, and spire is higher.

Cylichnella bidentata (d'Orbigny, 1841)

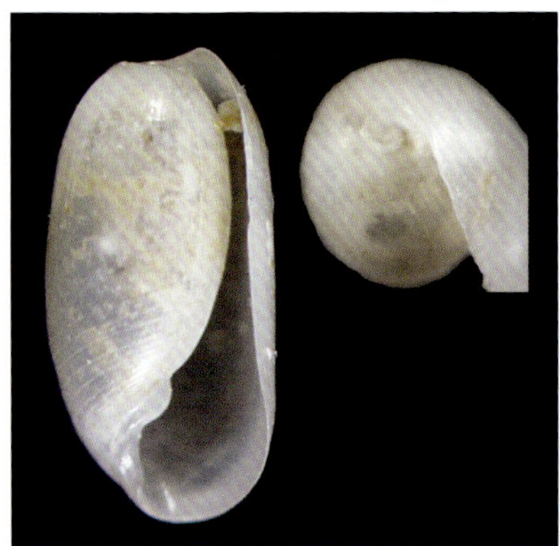

Two-toothed Barrel-bubble
Distribution: North Carolina, Florida, Texas; Brazil.
Size: 4 mm (⅙ in).

Description: Color translucent white. Shell shape elongate-cylindrical. Sculpture smoothish except for irregular axial growth lines and spiral threads found toward the base. Spire with slightly lowered callus engulfed by the rounded angle of the body whorl. Aperture elongate, narrow toward top, wider toward base; outer lip thin, lower lip rounded, and columella with strong, keel-like tooth. **Habitat:** In mud and sand; rarely in bays. Depth range 0 to 366 m (1200 ft). **Remarks:** Although spiral threads are typically seen at the base, there may be spirals toward the spire. **Synonym:** *Acteocina bidentata* (d'Orbigny, 1841).

Retusidae

Retusidae is a small family of gastropods comprising 3 genera and approximately 80 species. Retusids live in marine waters throughout the world and are found in the shallowest depths of bays and estuaries to bathyal depths greater than 5000 m (16,500 ft). Shells are pear shaped, oblong, or cylindrical. The aperture is long and broadens anteriorly. The protoconch may be sunken, short, and pointed. The posterior portion of the retusid shell is typically truncate. The outer surface is usually smooth, but axial sculpture, spiral sculpture, or a combination of both may be present. Retusids are known to feed upon foraminifera, other mollusks, and diatoms. Retusids and cylichnids are closely related but easily separated by the presence or absence of jaws and radulae. In Texas, Retusidae is represented by 2 genera and 5 species whose size ranges from 3 to 6 mm (⅛ to ¼ in); 3 species are discussed here.

Retusa caelatus (Bush, 1885)

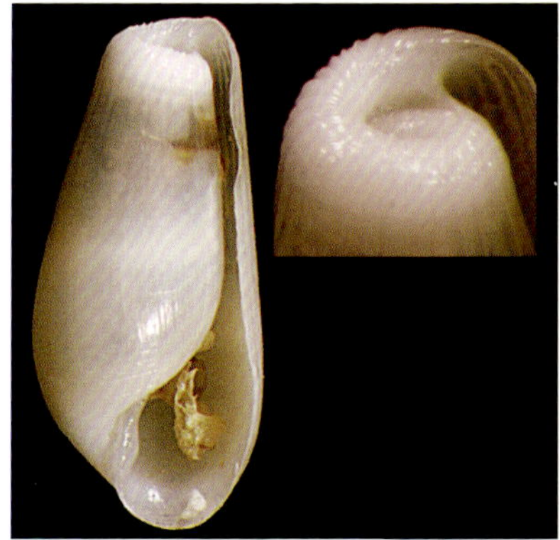

Engraved Barrel-bubble

Distribution: North Carolina, Florida, Texas; Campeche Bank, Mexico; Brazil.
Size: 4 mm (⅙ in).

Description: Color translucent, off white. Shell shape broad at one end and somewhat tapering at the other. Sculpture of axial ribs along the thickened spire portion of the shell in some specimens; in others axial lines travel the length of the shell. Protoconch sunken. Aperture elongate, narrow toward top, wider toward base; outer lip thin, and columella with a slight, toothlike protuberance at an angle. **Habitat:** Sandy bottoms at depths from 5 to 155 m (16 to 510 ft). **Remarks:** Quite a uniquely shaped micromollusk. **Synonym:** *Pyrunculus caelatus* (Bush, 1885).

Volvulella persimilis (Mörch, 1875)

Southern Spindle-bubble

Distribution: North Carolina, Florida, Texas-Louisiana; Yucatán; Brazil.
Size: 6 mm (¼ in).

Description: Color whitish. Shell shape oblong and cylindrical. Sculpture of spiral grooves on fresh shells, with the grooves barely detectable under magnification on worn shells. Protoconch pointed with a tiny grooved line and a channel-like projection. Aperture elongate, narrow toward top and wider toward base; outer lip strong, and columella with a cord, next to a chinklike umbilicus. **Habitat:** Sandy bottoms of the Texas-Louisiana coast. Depth range 0 to 805 m (2640 ft). **Remarks:** Has been found washed ashore in beach drift, but in Texas is found alive at depths from 15 to 42 m (51 to 138 ft).

Volvulella texasiana Harry, 1967

Texas Spindle-bubble

Distribution: Texas-Louisiana.
Size: 5 mm (⅕ in).

Description: Color translucent grayish-white, stained with reddish-brown at ends. Shell shape cylindrical. Sculpture smoothish. Spire with short, thin, and rounded spine on complete specimens. Aperture elongate, somewhat flared at top, narrow in middle, and flared toward base. Bottom area of lip arched, columella thickened, and joined with a short groove. **Habitat:** Collected along the Texas-Louisiana coast. Specimens have been collected at depths from 0 to 139 m (456 ft), but most living specimens are collected from depths between 15 and 51 m (48 and 168 ft). **Remarks:** The spine on the tip of the spire is typically broken off, which produces a bilobed appearance.

Cavoliniidae

Cavoliniidae is the most diverse and abundant family of gastropod mollusks that spend all their lives in the plankton. These mollusks have bilateral symmetry with extensive wing and mantle appendages. The cavoliniid shell is fragile, lacks coiling, and is extremely variable in shape. The identification of species within this family is primarily based on shell shape and number and shape of spines. Cavoliniids are opportunistic feeders and consume a variety of organisms, including diatoms, dinoflagellates, radiolarians, and foraminifera. Although planktonic, these pteropods are capable of locomotion by beating the highly flexible wing structures. On the Texas coast Cavoliniidae is represented by 5 genera and 9 species whose size ranges from 2 to 33 mm (1/12 to 1⅓ in); 2 species are discussed here.

Creseis acicula (Rang, 1828)   **Straight-needle Pteropod**

Distribution: 45°N to 42°S; 97°W to 0°W; Texas.
Size: Shell up to 33 mm (1⅓ in).

Description: Color translucent white. Shell shape needlelike. Sculpture smooth. Simple, narrow, tubelike shell, broader toward base and tapers to blunt point at apex. Shell straight or slightly curved. **Habitat:** Pelagic. Shells found at depths from 0 to 1234 m (4050 ft). **Remarks:** One of the most common pteropods found on Texas beaches. However, it is mostly found in pieces because of its fragile nature. The name *acicula* is appropriate because it means "small needle."

Diacavolinia longirostris (Blainville, 1821) **Long Snout Cavoline**

Distribution: 31.80°N to 39°S; 96°W to 38°W; Texas.
Size: 8 mm (⅓ in).

Description: Color translucent white. Shell shape globose with a tail-like projection. Sculpture weak on dorsal surface; sculpture on rounded ventral surface of irregular wavy ribs. Channel-like extension located centrally on back end and curved downward. Centrally located protrusion toward top, flanked by ear-like structures. **Habitat:** Pelagic. Shells found at depths from 0 to 1234 m (4050 ft). **Remarks:** This is one of the most abundant pteropods on the Texas coast. Usually found in greater numbers at depths from 36 to 72 m (120 to 240 ft). **Synonym:** _Cavolinia longirostris_ (Blainville, 1821).

Aplysiidae

The Aplysiidae is a small family of gastropods made up of about 60 species in 2 subfamilies and 9 genera, with 13 species found in the Gulf of Mexico. Their abundance in intertidal waters and large size make some members of the Aplysiidae the best-known opisthobranchs. The large head and "harelike" appearance of the rhinophores have led to their vernacular name, "sea hares." Mating is carried out in a chainlike structure, with the first aplysiid acting as a female and the last acting as a male; all in between act as both male and female. Coloration of aplysiids is determined by their plant-tissue diet. An internal shell may or may not be present. Locomotion is by a creeping action or by swimming, as in the genus _Aplysia_. Because of the easily accessible and large nervous system, _Aplysia_ have been used for extensive neurophysiological experiments. On the Texas coast Aplysiidae is represented by 2 genera and 5 species whose average size ranges from 65 to 400 mm (2½ to 15½ in); 4 species are discussed here.

Aplysia dactylomela Rang, 1828   Spotted Sea Hare

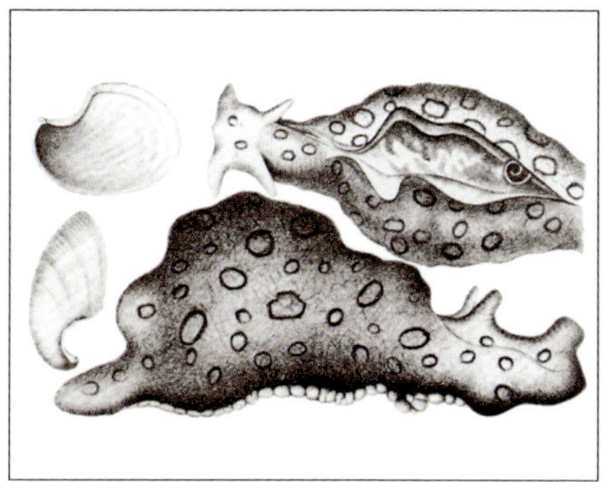

Distribution: Florida to Brazil; Texas; Bermuda; Africa.

Size: 410 mm (16 in) [body size, not shell, which is internal and reduced].

Description: Color usually greenish-brown with large black rings and black axial lines; black coloration inside and outside winglike structures. Animal shape winged, sluglike, with an ink sac located posteriorly on dorsal side. **Habitat:** Epifaunal. In algal mats. Depth range 0 to 3 m (10 ft). **Remarks:** Drawing from Andrews (1977). Usually found in Port Isabel in late fall and winter. **Synonym:** *A. protea* Rang, 1828.

Aplysia fasciata Poiret, 1789   Mottled Sea Hare

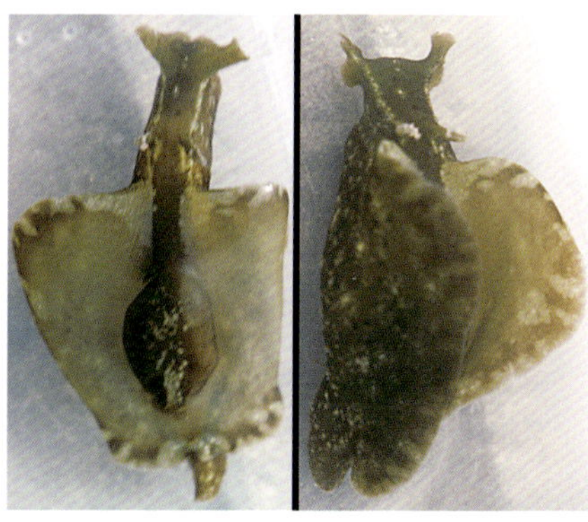

Distribution: Florida, Texas; Caribbean.

Size: 270 mm (10½ in) [body size, not shell, which is internal and reduced].

Description: Color variable from brown to light green with mottlings of lighter shades of green and brown, and spots of light yellow or white. Animal shape winged, sluglike, with an ink sac located posteriorly on dorsal side. **Habitat:** Cosmopolitan in temperate and tropical waters; inhabits protected waters such as bays and shallow seas. Depth range 0 to 13 m (43 ft). **Remarks:** Typically associated with abundant algae growth. **Synonyms:** *A. brasiliana* Rang, 1828; *A. willcoxi* Heilprin, 1887.

Aplysia morio (Verrill, 1901)   **Atlantic Black Sea Hare**

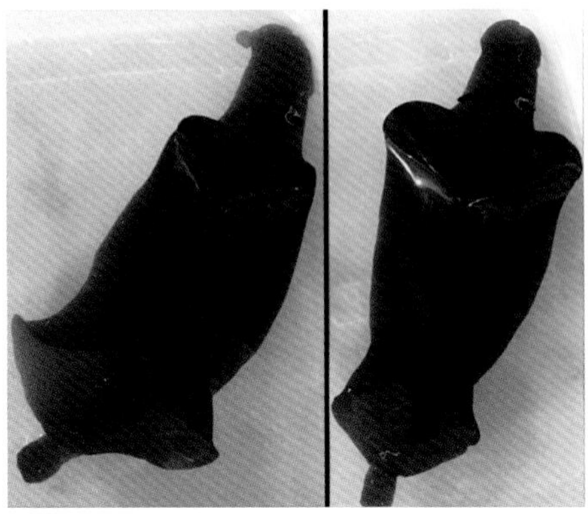

Distribution: Rhode Island, Florida, Texas; Bermuda; Caribbean.
Size: 400 mm (15½ in) [body size, not shell, which is internal and reduced].

Description: Color dark brown to black, mottling absent, may have darker lines on head and sides. Animal shape winged, sluglike, with an ink sac located posteriorly on dorsal side. Winglike structures free anteriorly, extending to the short, narrow posterior foot. Short tail. **Habitat:** Epifaunal. In algal mats. Depth range 0 to 42 m (140 ft). **Remarks:** Internal shell is long, slender, and ridged. **Synonym:** Mistakenly called _A. donca_ Marcus and Marcus, 1960.

Bursatella leachii Blainville, 1817   **Ragged Sea Hare**

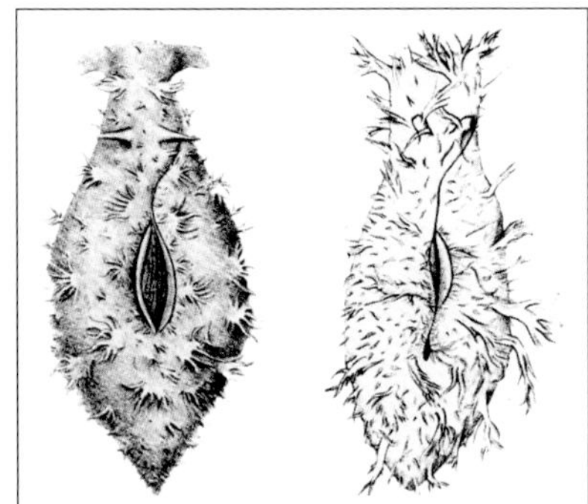

Distribution: Texas, Florida, Brazil.
Size: 120 mm (5 in) [body size, not shell, which is internal and reduced].

Description: Color variable black to brown, olive to gray with white, black, or gold splotches or spots. Animal shape globular with numerous irregular appendages extending from soft body. Camouflage appearance; difficult to determine external organs, such as head and eyes. **Habitat:** Epifaunal. In algal mats. Depth range 0 to 7 m (23 ft). **Remarks:** Drawings from Andrews (1977). Internal shell not present in adults. Several subspecies have been attributed to this genus (in Texas most commonly _B. leachii pleii_ Rang, 1828); however, some authors believe them to be just different forms of this species.

Scyllaeidae

Scyllaeidae is a small family of nudibranchs made up of 3 genera and a few species with only 1 species recognized in the Gulf of Mexico. All scyllaeids are capable of swimming to some extent. Scyllaeids have a dorso-lateral mantle ridge that exists as 1 or 2 flattened lobes on either side of the dorsal area. The rhinophoral sheaths (earlike tentacles) are equipped with a somewhat small rhinophoral club. The sheaths are flattened and rounded and may have a posterior keel. The rhinophoral club may possess a few horizontally placed lamellae and a terminally stalked knob. Posterior to the cerata (body outgrowths) there may be a medio-dorsal keel. Scyllaeids feed upon epiphytic hydroids that live on seagrasses and brown algae. On the Texas coast Scyllaeidae is represented by 1 species.

Scyllaea pelagica Linnaeus, 1758

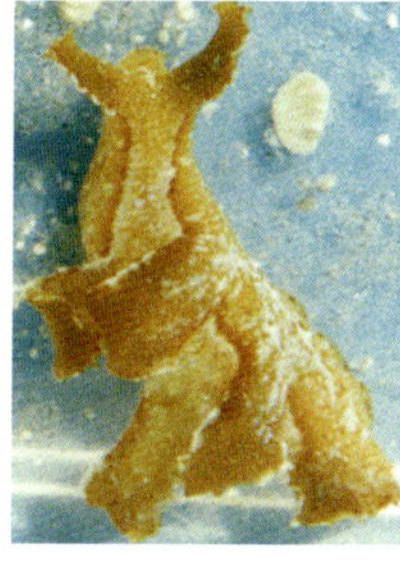 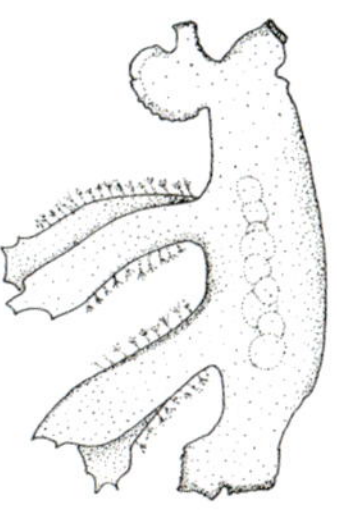

Sargassum Nudibranch

Distribution: Worldwide warm seas; Texas.

Size: 102 mm (4 in).

Description: Color variable, translucent green to orangish-brown with splotches of lighter shades to match their _Sargassum_ surroundings; may have row of blue spots on either side of body. Animal shape compressed with flattened, extended, elaborate, earlike structures and 2 pairs of clublike gill plumes. Pimplelike structures found scattered on sides of body. **Habitat:** Floating _Sargassum_. Depth range 0 to 11 m (36 ft). **Remarks:** Drawing from Andrews (1977). Animal feeds on hydroids attached to the _Sargassum_.

Glaucidae

At present the family Glaucidae is in a state of flux, and there is disagreement on the relationships among this group of nudibranchs. The glaucids are typically deep-bodied, slender nudibranchs with an elongate, tapering tail. They are small to relatively large and have been described as "pugnacious," or aggressive, animals. The oral tentacles are long and tapering, and a ridge usually separates the dorsal surface from the sides. The cerata are arranged in rows or cluster of rows or arches;

the rhinophores may be simple, ring-shaped, platelike, or have a nipple-like projection. The arrangement of the cerata has been used to distinguish subfamilies; however, the value of using these character sets may be doubtful because different ceratal arrangements may occur independently. Most glaucids feed on hydroids, but some may feed on soft corals or eggs of other opisthobranchs. On the Texas coast Glaucidae is represented by 1 species.

Glaucus atlanticus Forster, 1777

Blue Glaucus

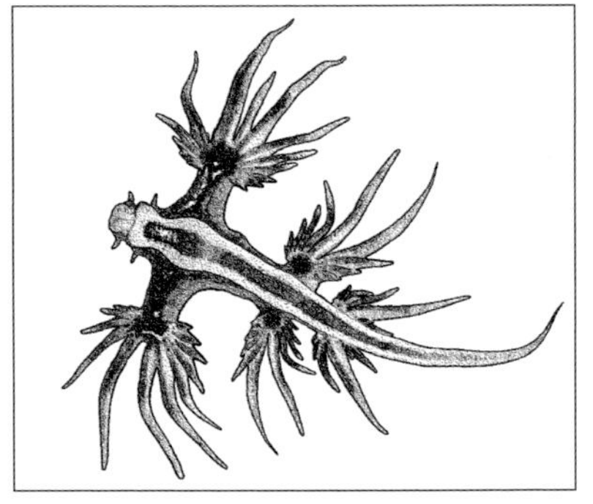

Distribution: Worldwide; Texas. **Size:** 25 mm (1 in).

Description: Color different shades of vibrant blues. Animal shape branched and sluglike. Oral tentacles and earlike structures small. Hornlike tentacles in frilled clumps on both sides of body. **Habitat:** Pelagic; warm seas. **Remarks:** Drawing by Moretzsohn in 2008. Feeds on planktonic-type cnidarians, such as the Portuguese man-of-war and by-the-wind sailor.

Ellobiidae

Ellobiidae is a relatively large family of gastropods that comprises approximately 20 genera and 300 species. These mollusks are typically located in tropical and temperate regions, living near the seacoast above the high-tide line. The shell is comparatively thick and usually has a high spire. Shell shape varies from long and conic, wide and oval, to subcircular. Sculpture may be smooth, spiral, axial, or both spiral and axial. The majority of the ellobiids have a monochrome coloration; however, some exhibit distinctive coloration. Opercula are absent. A modified lung has replaced the ancestral gill structures; and although they can breathe air, they are usually close to salt water. On the Texas coast Ellobiidae is represented by 2 genera and 2 species whose size ranges from 3 to 15 mm (⅛ to ⅗ in); both species are discussed here.

Melampus bidentatus Say, 1822   **Eastern Melampus**

Distribution: Quebec to Texas; Caribbean; Bermuda.

Size: 20 mm (⅘ in).

Description: Color variable, usually brownish with white bands. Shell shape ovate. Sculpture smoothish. Body whorl dominant portion of shell. Spire with pimplelike protoconch. Aperture long and narrow, inner area of outer lip with toothlike structures, and columella may have 2 or 3 columellar folds. **Habitat:** In bays and lagoons under vegetation, above tide line. **Remarks:** This species varies in coloration, size, and number of denticles on inner area of outer lip and number of folds on columella.

Pedipes mirabilis (Mühlfeld, 1816)   **Miraculous Pedipes**

Distribution: Florida, Texas; Bermuda; Brazil.

Size: 6 mm (¼ in).

Description: Color of exterior brown; interior white. Shell shape globose. Sculpture of somewhat shouldered convex whorls. Body whorl about ¾ shell length with slanting, wavy, irregular axial lines crossed by spiral threads, giving latticelike appearance. Protoconch somewhat sunken, smooth. Aperture subcircular, strongly ornamented, partially exposed by white, strong, irregular toothlike structures on parietal shield, extending to inner area of outer lip. **Habitat:** Epifaunal. Hard substrate near high-tide line to a depth of 1.5 m (5 ft). **Remarks:** Shells found in beach drift.

Siphonariidae

Siphonariidae is a family of gastropods, also known as false limpets, made up of 3 genera and about 75 species. Siphonariids typically inhabit shallow tidal areas from subantarctic regions to temperate and tropical regions worldwide, with the exception of the North Atlantic. These gastropods are air breathers and live in a semi-marine environment. Their flattened, cone-shaped shell resembles that of the true limpets, but false limpets lack true gills. Instead, either they have secondary gills within the mantle cavity or the mantle cavity can serve both as a lung and a gill. The shell of siphonariids is ovate and asymmetrical and has a subcentral apex that is often eroded. The shell is flattened conic, is oval to round in appearance, and may be either smooth or radially sculptured. False limpets feed primarily on microscopic plant life. On the Texas coast Siphonariidae is represented by 2 genera and 2 species whose average size ranges from 4 to 25 mm (⅙ to 1 in); 1 species is discussed here.

Siphonaria pectinata (Linnaeus, 1758)

Striped False Limpet

Distribution: Florida, Texas; Mexico; Caribbean.

Size: 27 mm (1 in).

Description: Color of numerous closely spaced white ridges, separated by many crowded brownish grooves that separate into 2 branches toward the margin; inner area glossy with brownish coloration bleeding through. Shell shape limpetlike. Sculpture of many crowded, alternating radial ridges and grooves, with margin finely crenulate. Aperture subcircular. **Habitat:** Intertidal. Epifaunal on algae-covered rocks and other hard substrates. **Remarks:** Common on Texas coast jetties. The shell resembles a true limpet but does not have a hole on the top of the caplike shell.

PART III

CLASS: CEPHALOPODA
Squids and Octopuses

The class Cephalopoda is made up of about 900 species and 2 subclasses, Nautiloidea and Coleoidea. Nautiloids are chambered and coiled; they lack suckers, and *Nautilus* is the only extant genus with an outer shell. Coleoids begin with an inner shell that is enclosed in a sac that will either develop a shell, inhibit the shell, or suppress the shell entirely. Cephalopods are the most highly evolved of the 7 molluscan classes. Cephalopods live in all oceanic environments from the Antarctic to the Arctic. Their mode of living resembles fishlike activities rather than those of other classes of mollusks. Evolutionary development of bilateral symmetry within the cephalopods has remained consistent with that of their ancestral counterparts. Cephalopods have strong, beaklike jaws that enable them to prey on large animals. In Texas, Cephalopoda is represented by 6 families, 7 genera, and 10 species; 3 families and 4 species are discussed here. However, confirmed sightings of some of these secretive species are rare.

Spirulidae

Spirulidae is a family of small mollusks that rarely exceed 45 mm (1⅘ in) in length. The internal, spirally coiled shell is located posteriorly in the animal. The adult shell is made up of about 30 chambers. Spirulids possess a stout but thin mantle, and the cartilage locking system of funnel and mantle is plain and straight. This family is recognized by the unique internal structure of its shell, which is shaped like the arched horns of a ram. Spirulids are found in the open ocean, mostly at depths of about 400 m (1320 ft) where water temperature is about 10°C (50°F) or warmer. Living spirulids take a body position in the water column with their head facing downward and use jet propulsion as the means of locomotion. In Texas, Spirulidae is represented by 1 species.

Spirula spirula (Linnaeus, 1758) **Ram's Horn Squid**

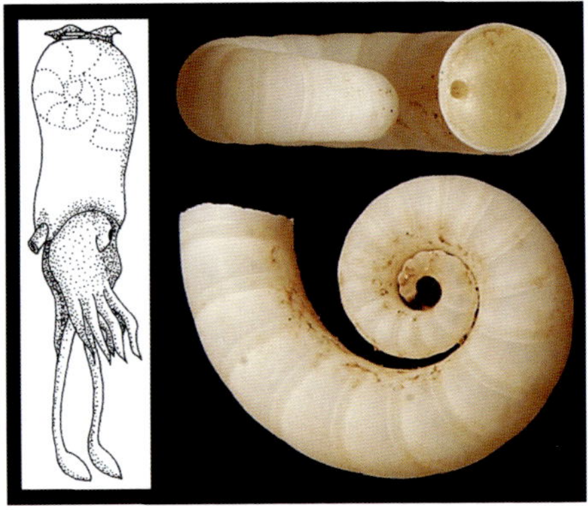

Distribution: Worldwide; Texas.
Size: 25 mm (1 in).

Description: Color of inner shell yellowish-white. Shape coiled. Sculpture smoothish, except for circular spiral rings that separate the chambers. Aperture circular, covered except for a small hole that leads into the next chamber. Animal has 10 appendages, 8 arms, and 2 tentacles. **Habitat:** All oceans. **Remarks:** Drawing from Andrews (1977). Rarely seen alive; however, its small internal shell is found washed ashore on beaches worldwide. Alive at depths from 550 to 1000 m (1804 to 3300 ft) where water temperature is about 10°C (50°F) or warmer.

Loliginidae

Loliginidae is represented by 11 genera. The 10 appendages, transparent cornea covering the eyes, straight and simple mantle locking mechanism, 7 buccal lappets, and buccal connectives that attach to the ventral borders of the ventral arms are major characteristics that distinguish the loliginids from other cephalopods. Mature males typically use the left ventral arm for mating. Courtship precedes mating. The male pursues the female as he displays his courtship colors, and the female, if agreeable, displays her colors and mating occurs. Jet propulsion is the primary means of locomotion. Loliginids are important in fisheries as a food source, but they are also highly prized for the neurophysiological importance of their giant axons, which are commonly used in laboratory experiments. In Texas, Loliginidae is represented by 3 species whose size ranges from 127 to 610 mm (5 to 24 in); 2 species are discussed here.

Doryteuthis plei (Blainville, 1823)

Slender Inshore Squid

Distribution: Worldwide; Texas. **Size:** 203 mm (8 in).

Description: Color of live specimens with reddish-purple spots covering the surface of the mantle. Shape fusiform. Mantle long, narrow, tapering posteriorly. Tentacular arms long, equipped with tentacular clubs with 4 rows of suckers on each club; tentacles with 2 rows of suckers. Triangular fins on last ⅓ of mantle. **Habitat:** Pelagic. At depths from just below surface to 366 m (1208 ft). **Remarks:** Photograph by NOAA. **Synonym:** *Loligo plei* (Blainville, 1823).

Lolliguncula brevis (Blainville, 1823)   **Atlantic Brief Squid**

Distribution: Maryland, Texas; Brazil; Bermuda.
Size: 127 to 254 mm (5 to 10 in).

Description: Color reddish-orange spots over yellowish-orange base dorsally, ventral surface yellowish-white with reddish-orange spots on mantle, ventral surface of fins whitish-yellow with no spots. Shape short and stocky; width of posterior end broad, not tapered; anterior end narrower. Fins together wider than long and do not reach the center of the length of the mantle. Internal shell visible through translucent mantle. **Habitat:** Shallow water from just below surface to 18 m (60 ft). **Remarks:** *Doryteuthis plei* and *D. pealeii* mantles taper toward the posterior end of the mantle, but mantle of *Lolliguncula brevis* does not. **Synonym:** *Loligo brevis* (Blainville, 1823).

Octopodidae

Octopodidae is a group of benthic cephalopods. Their bodies are muscular, and their 8 arms are equipped with either 1 or 2 rows of suckers that lack chitinous rings. The arms are connected by weblike material that can be superficial or rather deep. Also lacking is the mantle funnel locking mechanism. Locomotion in octopodids is usually by crawling across the substrate, which is efficient because of their muscular body and powerful suckers. They can also move by jet propulsion, but it is not as effective as in other cephalopods because of the octopodid globular form and lack of fins. Arms in this family are numbered in a specific manner and are used for identification purposes, and the genus *Octopus* includes only 8 known species. The octopodid reproductive scheme is their undoing, as they die after one season. Octopuses have many predators, such as bottlenosed dolphins, pilot whales, fur seals, and hammerhead sharks. Diets of octopodids vary and include polychaetes, fishes, crustaceans, and other mollusks. In Texas Octopodidae is represented by 1 genus and 3 species whose size ranges from 102 to 914 mm (4 to 36 in); only the most common species is discussed here.

Octopus cf. *vulgaris*   Common Octopus

Distribution: Connecticut, Florida, Texas; Cuba.
Size: 305 to 914 mm (1 to 3 ft), including tentacles.

Description: Color reddish-brown. Body saclike. Mantle width about 7/10 mantle length. Head smallish. Arms long. Skin typically smooth in life and wrinkled when preserved. Hectocotylized arm is short. **Habitat:** All bottom types from near shore to edge of continental shelf. Found at depths from below surface to 200 m (660 ft). **Remarks:** This is now considered a species complex known as *Octopus* cf. *vulgaris* group. In Texas found on jetties and offshore reefs and banks; males usually die after mating, and females perish after eggs are hatched.

PART IV

CLASS: BIVALVIA
Oysters, Mussels, and Clams

The class Bivalvia includes mussels, oysters, scallops, and clams. This class of mollusks is named for the 2-shelled valves that protect the soft parts of the animal. Bivalves are an important part of the benthic community and reside from the surface to the abyss. They can be found attached to or boring into hard substrates, such as rocks, shells, and wood; buried in mud and sand; or living on seagrass. They serve as filter feeders that ingest organic material taken from the water column and discharge organic by-products to the substrate. Burrowing species turn over soft substrate, thereby aerating the topmost substratum. Boring species break down coral rock, wooden materials, and other hard organic material, thus creating habitat for other biota. Other bivalves, such as shipworms, are destructive because they burrow into and destroy wooden pilings and other structures. Edible bivalves, such as scallops, oysters, mussels, and clams, have great commercial value. Most bivalves are free living, although some are parasitic. Some bivalve species, such as oysters, have 1 valve cemented to the substrate and are not bilaterally symmetrical. Bivalves are a morphologically diverse group of mollusks that are clearly related to one another. All bivalves have, as their earliest larval stage, a trocophore-type (a free-swimming planktonic) larval stage. A total of 275 species in 61 families have been reported from the Texas coast; in this field guide the most common 137 species in 40 families are discussed and illustrated.

Nuculidae

Nuculidae is a medium- to large-sized family of bivalves, commonly called nut shells, made up of about 150 species. The color varies from pale yellow to brown in fresh material. Most outer shell sculpture is smooth but may have microscopic commarginal lines and/or radial threads. Shell shape is from ovate to triangular, inflated with nacreous inner layers. Beaks are positioned posteriorly and point toward the back, with a concave ligamental shelf located beneath the beak. The hinge area has a series of V-shaped, erect teeth separated by a spoon-shaped object (chondrophore). The pallial line is simple but obscure, and the pallial sinus is absent. Nuculids live by burrowing into the superficial layers of sediment and inhabit areas just below the surface. The foot of these bivalves is made of 2 lateral flaps that enable the animal to anchor itself and to advance by contraction of pedal retractor muscles. In Texas, Nuculidae is represented by 2 genera and 6 species whose size ranges from 1.9 to 18 mm (¹⁄₁₂ to ¾ in); 1 species is discussed here.

Nucula proxima Say, 1822 — **Atlantic Nut Clam**

Distribution: Canada; Florida, Texas, Gulf of Mexico; Bermuda; Caribbean.

Size: 3 to 10 mm (⅛ to ⅖ in).

Description: Color whitish to gray, may have a purplish band just below the umbo in fresh specimens. Shape somewhat triangular. Sculpture smooth with distinct darker-shaded commarginal growth lines. Interior glossy, whitish-gray with fine radial threads showing at the surface; interior ventral margin with minute, somewhat beaded, denticulations. **Habitat:** Infaunal. Sandy mud-type substrate at depths from 0 to 124 m (407 ft). **Remarks:** Specimen in photograph from HMNS collection; listed as *N. aegeensis.*

Nuculanidae

Nuculanidae is a large family of bivalves of about 220 species commonly known as elongate nut clams. Unlike their relatives, the Nuculidae, these bivalves do not have pearly or nacreous inner shell layers. Shells are thin, and the posterior end typically tapers; the anterior end is rounded. Valves of the nuculanids are typically smooth but may have concentric growth lines, and radial ribs may be present. The hinge line has chevron-type teeth separated by a spoon-shaped object (chondrophore). The pallial line is somewhat impressed, and the pallial sinus may be small or large. Rapid burrowing is accomplished by the presence of 2 lateral flaps that make up the foot. Nuculanids use their proboscis, along with the pumping action and filtration process of the gill structures, to feed on detritus. The lower taxonomic status of extant nuculanids is in a state of confusion, as is the case with many small mollusks. In Texas, Nuculanidae is represented by 1 genus and 4 species whose size ranges from 6 to 18 mm (¼ to ¾ in); 2 species are discussed here.

Nuculana acuta (Conrad, 1832) — **Pointed Nut Clam**

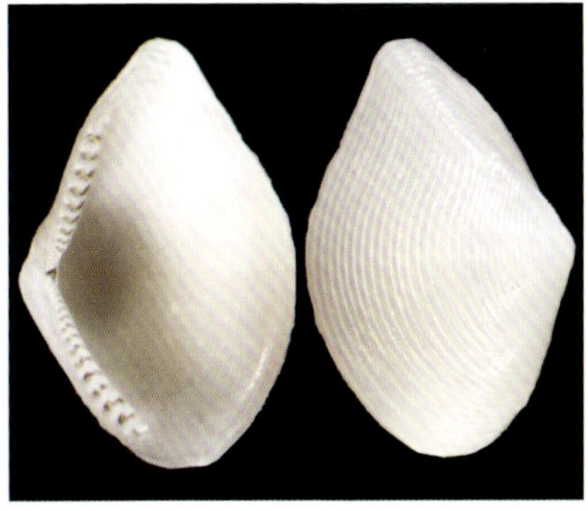

Distribution: Massachusetts to Texas; Caribbean; Brazil.
Size: 6 to 9 mm (¼ to 3/8 in).

Description: Color whitish. Outline broad and elongate. Sculpture of numerous, regular, concentric, modular ridges with a ridge on the posterior dorsal margin. Distinct escutcheon and lunule are present. Triangular chondrophore with numerous teeth found on either side.
Habitat: Sandy, shelly bottoms. Depth range from 0 to 274 m (900 ft).
Remarks: Dead shells are often found at Stetson Bank at depths from 21 to 30 m (70 to 100 ft). Usually taken alive at depths from 36 to 137 m (120 to 456 ft). Similar to *N. unca,* but broader and with finer concentric threads.

Nuculana concentrica (Say, 1824) **Concentric Nut Clam**

Distribution: Florida to Texas; Brazil.

Size: 12 to 18 mm (½ to ¾ in).

Description: Color pale yellow to white, semi-glossy. Shape elongate. Sculpture smooth with fine, irregular commarginal growth lines; smooth radial ridge from umbones to posterior edge. Pallial line impressed; pallial sinus relatively small, rounded anteriorly, pointed posteriorly. Dentition on both sides of umbo, chevron shaped.
Habitat: Infaunal in sandy mud-type sediment in bays and offshore. Along the entire Texas coast; subtidally from shoreline to a depth of 90 m (295 ft). **Synonym:** *N. eborea* Conrad, 1848.

Arcidae

Arcidae is a diverse family of marine bivalves called ark shells. It is a moderate to large family of about 140 species. These bivalves are located in marine environments of temperate and tropical climates and are usually collected intertidally in shallow, sublittoral zones. Shells vary in outline. Radial sculpture typically dominates; concentric sculpture can be strong but usually is of commarginal growth lines. Dentition is taxodont (small parallel to subparallel teeth that are perpendicular to hinge line), and teeth are small to moderate-sized. The hinge line is either straight or weakly arched. Ark shells typically have 2 subequal, roundish muscle scars, one on either side of the shell. Beaks are prominent. Shell characters defining this family are taxodont dentition; elongate duplivincular (dorsal ligament made up of a series of bands attached to a shallow groove) hinge ligament; and a flattened, dorsal, central region that separates the beaks. Most arcids are byssate (silky threads secreted to attach the bivalve to hard substrates) nestlers on hard substratum, such as coral and rocks. Others are found burrowing in mud or sand and can be collected in great numbers under suitable conditions. In Texas, Arcidae is represented by 10 genera and 15 species whose size ranges from 4 to 102 mm (⅙ to 4 in); 12 species are discussed here.

Acar domingensis (Lamarck, 1819)   **White Miniature Ark**

Distribution: North Carolina to Texas; Brazil; Bermuda.
Size: 12 to 30 mm (½ to 1⅕ in).

Description: Color uniformly white. Shape rhomboidal. Sculpture strong, with radial ribs crossed by strong concentric ribs, giving sturdy, cancellate appearance; ridge or sulcus present from umbones to posterior ventral angle; cancellate sculpture seen through shell. Hinge line straight with numerous teeth that become smaller toward central area. **Habitat:** On hard surfaces. Epifaunal; byssally attached. **Remarks:** Found in beach drift from Sargent Beach south, alive at South Padre Island, in depths from below tide line to 155 m (510 ft). **Synonym:** _Barbatia domingensis_ Lamarck, 1819.

Anadara baughmani Hertlein, 1951   **Baughman's Ark**

Distribution: Texas, Gulf of Mexico; Brazil.
Size: 38 to 54 mm (1½ to 2⅛ in).

Description: Color uniformly white. Shell thick, ovate, and inflated. Sculpture of thick, radial, modular ribs and commarginal growth lines, giving a somewhat scaly appearance; numerous small concentric riblets between radial ribs. Hinge plate long with vertical comblike teeth that become smaller toward the center. Ligamental area wide and short. **Habitat:** Muddy bottoms. Infaunal. On Texas coast at depths from 0 to 183 m (600 ft). **Remarks:** Common name listed as "Baughman's ark." Turgeon et al. (1998) suggest "skewed ark" to describe anterior margin.

Anadara floridana (Conrad, 1869) **Cut-ribbed Ark**

Distribution: North Carolina to Texas; Greater Antilles. **Size:** 63 to 102 mm (2½ to 4 in).

Description: Color white. Shape obliquely rectangular. Sculpture of modular radial ribs with fine concentric riblets crossing over radial ribs. Hinge long with numerous comblike vertical teeth that become smaller toward the center. Ligamental area long and narrow; pallial line simple. **Habitat:** Beaches along entire Texas coast. Depth range from 0 to 110 m (360 ft). **Remarks:** Found alive on South Padre Island. Living, large-sized animals are rare but are probably located at depths from 14 to 47 m (48 to 156 ft).

Anadara transversa (Say, 1822) **Transverse Ark**

Distribution: Cape Cod, Massachusetts to Florida, Texas, and Campeche Bank, Mexico. **Size:** 12 to 42 mm (½ to 1¾ in).

Description: Color yellowish-white; interior white. Outline variable, typically oblong. Sculpture of radial modular ribs that are typically beaded on left valve but not right; concentric sculpture of commarginal growth lines. Hinge line straight with numerous erect teeth. Ligament area long and narrow; ventral margin serrate. **Habitat:** Infaunal in sandy bottoms or epifaunal on hard substrates (especially when young). Found at depths from shoreline to 73 m (240 ft). **Remarks:** Abundant in bays and beach drift on entire Texas coast.

Arca imbricata Bruguière, 1789   **Mossy Ark**

Distribution: North Carolina to Texas; Caribbean; Brazil; Bermuda.

Size: 38 to 63 mm (1½ to 2½ in).

Description: Color light brown to whitish at umbones; interior whitish-purple. Shape elongate-rectangular with posterior end broader. Sculpture of radial ribs crossed by commarginal growth lines, giving beaded, cancellate appearance; a sturdy ridge or sulcus extends from umbo to posterior margin. Hinge line long, narrow, and straight with numerous erect teeth that become smaller toward the center. **Habitat:** Byssally attached to hard substrates; epifaunal. Found at depths from shoreline to 64 m (210 ft). **Remarks:** Live at South Padre Island, Seven and One-half Fathom Reef, and on artificial reefs of Texas.

Arca zebra (Swainson, 1833)   **Turkey Wing**

Distribution: North Carolina to Texas; Caribbean; Brazil; Bermuda.

Size: 50 to 84 mm (2 to 3¼ in).

Description: Color yellowish-white with wavy commarginal reddish-brown bands; interior yellowish-white in center with brownish-purplish bands toward margins. Shape elongate-rectangular. Sculpture of modular radial ribs with commarginal growth lines that are thicker at ventral margin, giving scaly appearance; angled sulcus from umbones to posterior ventral margin. **Habitat:** Shelly bottoms and calcareous environments; epifaunal. Found at depths from shoreline to 140 m (460 ft). **Remarks:** On jetties at Brazos Santiago Pass.

Barbatia cancellaria (Lamarck, 1819)   **Red-brown Ark**

Distribution: Florida, Texas; Caribbean; Brazil.
Size: 25 to 30 mm (1 to 1⅕ in).

Description: Color yellowish-white, sometimes with brown maculations when worn, reddish-brown when fresh; interior yellowish toward center, whitish toward margins. Shape obliquely rectangular. Sculpture of thin, furrowed radial ribs and fine, obscure, commarginal growth lines. Hinge narrow and straight with teeth oblique, inclined toward central area. **Habitat:** Calcareous environments and rocky shores. Found occasionally on beaches along the Texas coast and at depths from 15 to 85 m (48 to 280 ft). Depth range from 0 to 85 m (280 ft).

Cucullaearca candida (Helbling, 1779)   **White Bearded Ark**

Distribution: North Carolina to Texas; Brazil.
Size: 38 to 80 mm (1½ to 3⅛ in).

Description: Color uniformly white. Shape oblique, somewhat rectangular. Sculpture of numerous, compact, modular radiating ribs with regularly spaced knobs on adjacent ribs, giving beaded appearance; anterior end rounded, posterior end flared; distinct ridge from umbo to posterior ventral angle. Hinge long, straight, and narrow with oblique teeth inclined toward central area. **Habitat:** Beyond low-tide line on hard substrates; epifaunal byssal nestler. Entire Texas coast from below tide line to a depth of about 101 m (330 ft). **Remarks:** Found alive at South Padre Island. **Synonym:** _Barbatia candida_ (Gray, 1842).

Fugleria tenera (C. B. Adams, 1845)

Distribution: Florida, Texas; Caribbean; northern coast of South America.
Size: 25.0 to 33.5 mm (1 to 1⅖ in).

Description: Color white; interior yellowish with whitish margins. Shape oblong-quadrate, inflated at umbones. Sculpture of slender, alternating, large and small, threadlike radial ribs with concentric growth lines, giving cancellate appearance; numerous riblets between radial ribs. Hinge line straight; teeth chevron shaped; ligament area relatively wide at umbo. **Habitat:** Attached by byssus to hard surfaces; epifaunal nestler. Entire Texas coast from shoreline to a depth of approximately 46 m (150 ft). **Synonym:** *Barbatia tenera* (C. B. Adams, 1845).

Lunarca ovalis (Bruguière, 1789)

Distribution: Cape Cod, Massachusetts to Texas; Caribbean; Brazil.
Size: 38 to 59 mm (1½ to 2⅓ in).

Description: Color uniformly white with a persistent brown periostracum. Shape subcircular, inflated. Sculpture of exteriorly flattened, grooved radial ribs that get wider toward the ventral margin; concentric riblets between radial ribs give cancellate appearance at umbones; ventral margin serrate. Ligamental area thin; hinge line slightly arched. **Habitat:** Mud and sandy bottoms in shallow bays and just offshore. Associated with *Noetia ponderosa* at depths from shoreline to 68 m (223 ft). **Remarks:** Unlike most mollusks, this bivalve has red blood or hemoglobin, thus the common name "blood ark." **Synonym:** *Anadara ovalis* (Bruguière, 1789).

Scapharca brasiliana (Lamarck, 1819)   **Incongruous Ark**

Distribution: North Carolina to Florida to Texas; Brazil.
Size: 25 to 62 mm (1 to 2½ in).

Description: Color whitish with brown periostracum; interior yellowish-white. Shape inflated, broadly ovate. Sculpture of broad, flattened, horizontally wide, vertically narrow, beaded radial ribs with largest beads toward ventral margin; area between ribs smooth. Hinge with comblike teeth becoming smaller toward umbones; ligamental area wide and short; ventral inner margin distinctly serrate; inward-pointing umbones. **Habitat:** Infaunal. Sandy areas along shoreline. On Texas coast from shoreline to a depth of 75 m (246 ft). **Synonym:** _Anadara brasiliana_ (Lamarck, 1819); _A. incongrua_ (Say, 1822).

Scapharca chemnitzii (Philippi, 1851)   **Chemnitz's Triangular Ark**

Distribution: Texas; Caribbean to Brazil.
Size: 28 mm (1⅛ in).

Description: Color white. Shape triangular-oval. Sculpture similar to that of _A. brasiliana_ but radial ribs thinner, beads not as elongate, and numerous concentric ribs in area between radial ribs. Pallial line simple. Hinge line short with many comblike teeth becoming smaller toward the center. **Habitat:** Commonly found in beach drift south of St. Joseph Island. Found on entire Texas coast from shoreline to a depth of about 75 m (246 ft). **Remarks:** Resembles _A. brasiliana_ when fully mature. Common name listed as "Chemnitz's ark." Turgeon et al. (1998) suggest "Chemnitz's triangular ark" to describe its shape. **Synonym:** _Anadara chemnitzii_ (Philippi, 1851).

Noetiidae

Members of the family Noetiidae are similar to the arcids but are differentiated by the ligament being confined to only part of the cardinal region of the shell. Noetids possess vertically arranged taxodont dentition and straight or slightly arched hinge lines. Shell composition is much like that of the Arcidae. Shell shape is from ovate to elongate. The shell is typically equivalve (2 valves of equal size), strong, and inflated. The size of these bivalves is small (4 to 5 mm, or ⅙ to ⅕ in) to medium (63 mm, or 2½ in). Sculpture is radial or cancellate. Mantle margins are muscular and possess eyespots on the dorsal anterior region. In some genera the posterior muscle scar can be large; the anterior scar can be small; and the anterior muscle scars can both be small in some and almost equal in others. The periostracum is brown, dense, and fibrous. Much confusion exists at the subfamilial and generic levels of this family, and much more work is needed to correct the taxonomy of the Noetiidae. In Texas Noetiidae is represented by 2 genera and 2 species whose size ranges from 6 to 63 mm (¼ to 2½ in); both species are discussed here.

Arcopsis adamsi (Dall, 1886)

Adams Miniature Ark

Distribution: North Carolina to Texas; Caribbean; Brazil; Bermuda.

Size: 6 to 14 mm (¼ to ½ in).

Description: Color yellowish; interior white. Shape inflated, rectangular. Sculpture thick with radial ribs crossed by concentric ribs, giving reticulated pattern. Umbones large; ligament external, triangular, and restricted to area beneath umbones; hinge line straight with many small teeth on both sides of umbones; no teeth beneath umbones. **Habitat:** Byssally attached to undersides of hard surfaces; epifaunal. Depth range from 0 to 128 m (420 ft). **Remarks:** In Texas found at the Flower Garden Banks. Common name listed as "Adams miniature ark." Turgeon et al. (1998) suggest "cancellate ark" to describe its sculpture.

Noetia ponderosa* (Say, 1822)   **Ponderous Ark**

Distribution: Virginia to Florida, Texas.

Size: 51 to 76 mm (2 to 3 in).

Description: Color uniformly white; hirsute brown periostracum. Shape inflated, thick, trapezoidal, widely ovate. Sculpture of flattened, grooved radial ribs with fine concentric threads between radials. Hinge line straight; ligament area relatively broad with numerous, various-sized teeth, anterior teeth angled. Inner ventral margin serrate. Posterior adductor muscle scar raised to form a rim. **Habitat:** Coastal bays and sandy bottoms offshore at depths from below tide line to 68 m (223 ft). **Remarks:** This is one of the most widespread bivalves on the entire Texas coast.

Mytilidae

Mytilidae, a somewhat large family of bivalves that is subdivided into 6 subfamilies and approximately 250 species, includes some of the most conspicuous and gregarious bivalves along a marine shore. Shell shape may be ovate, elongate, or fanlike. Anteriorly turned beaks are located at the narrower end. The posterior end is rounder and wider. Sculpture may be smooth, axial, or cancellate. Dentition is dysodont (hinge that bears small simple denticles close to the umbones). The interior surface of the shell has 2 muscle scars: a small one toward the front (narrow end) and a larger one toward the rear. Juveniles are able to move by removing byssal threads; however, when a significant amount of byssal thread is laid down, adults become sedentary and subsequent movement is only by the mytilid reorienting itself on the byssal mass. Mytilids are suspension feeders living by attaching to a firm substrate and existing either singly or in bunches. In Texas, Mytilidae is represented by 13 genera and 20 species whose size ranges from 6 to 180 mm (¼ to 6 in); 12 species are discussed here.

Amygdalum papyrium (Conrad, 1846)   **Atlantic Paper Mussel**

Distribution: Maryland to Florida, Gulf of Mexico, Texas; Brazil.

Size: 18 to 22 mm (¾ to 5/6 in).

Description: Color white with blue-green to tannish periostracum; interior iridescent. Shape ovate and long, wider posteriorly. Sculpture smooth. Shell fragile with commarginal growth lines. Hinge ligament on a thin, obscure shelf just below the umbones. **Habitat:** In Texas found nestling in seagrass roots and abandoned shells. Depth range from 0 to 38 m (125 ft). **Remarks:** Found on entire Texas coast, along most bay margins. Environment and type of substrate play a significant role in color and size of these bivalves.

Botula fusca (Gmelin, 1791)   **Cinnamon Mussel**

Distribution: North Carolina to Florida, Texas; Caribbean; Bermuda; Brazil.

Size: 18 to 25 mm (¾ to 1 in).

Description: Color brownish. Shape broadly elongate, robust. Sculpture smooth with thick, irregularly spaced commarginal growth lines. Umbo rolled inward posteriorly; numerous vertical striae positioned anterior to the hinge ligament. **Habitat:** In Texas found at the Flower Garden Banks, in calcareous environments, burrowed in hard substrate. Depth range from 2 to 93 m (7 to 305 ft). **Remarks:** Easily confused with *Lioberus castaneus* (Say, 1822). Many specimens exceed the 18 mm (¾ in) size listed in Abbott (1974), with maximum size listed at 28 mm (1⅛ in).

Brachidontes exustus (Linnaeus, 1758)   **Scorched Mussel**

Distribution: North Carolina to Texas; Caribbean; Brazil to Uruguay.

Size: 18 mm (¾ in).

Description: Color yellow-brown with darker commarginal bands; interior metallic purple with splotches of white. Shape triangular. Sculpture of numerous rounded, interrupted radial ribs. Beak slightly rolled inward; interior posterior marginal edge with teeth that progressively increase in size from umbo to ventral margin; ventral margin with crenulations made by external radial ribs; antero-marginal edge slightly arched. **Habitat:** Attached to hard substrate in brackish water at depths from 0 to 46 m (150 ft). **Remarks:** Differs from *B. domingensis* in having a deeper body cavity and finer and more numerous ribs.

Geukensia granosissima (Sowerby III, 1914)   **Southern Ribbed Mussel**

Distribution: Gulf of St. Lawrence to Florida, Texas.

Size: 50 to 100 mm (2 to 4 in).

Description: Color brown to black; interior iridescent bluish-white with ventral margin purplish. Shape long and narrow. Sculpture with bifurcating ribs crossed by small riblets and larger commarginal growth lines; outer margin smooth; inner anterior margin with thin hinge line below umbo; posterior inner margin thin and smooth; ventral margin crenulate. **Habitat:** Intertidal zone of salt marshes; root systems of *Spartina alterniflora* along the bay margins of Galveston and Matagorda Bays. **Remarks:** Formerly recognized as *G. demissa* (Dillwyn, 1817); however, *G. demissa* has a more northerly range. These mussels are important in the flow of energy through a salt marsh.

Gregariella coralliophaga (Gmelin, 1791)   **Coral-eating Mussel**

Distribution: North Carolina to Texas; Caribbean; Bermuda; Brazil.

Size: 18 mm (¾ in).

Description: Color of exterior white, rarely with purplish-bluish stain. Shape robust, broad, rectangularly elongate. Sculpture cancellate with periostracum mainly covering umbonal keel. Umbo located at extreme anterior end and rolled inward; lateral teeth on both sides of umbones, prominent on lower portion, no cardinal teeth; thin hinge line. **Habitat:** Vacated shells and old bore holes, mainly offshore. Depth range from 0 to 101 m (330 ft). **Remarks:** There is some question as to the differences between *G. coralliophaga* and *G. opifex*. Abbott (1974) considers *G. opifex* a synonym of *G. coralliophaga*. See Tunnell et al. (2010: 315–316) for more details on these 2 species.

Ischadium recurvum (Rafinesque, 1820)   **Hooked Mussel**

Distribution: Cape Cod, Massachuetts to Caribbean; Texas.

Size: 25 to 63 mm (1 to 2½ in).

Description: Color bluish-brownish with streaks of white; interior dark purplish-brown with whitish margins. Shape elongate and relatively narrow and curved. Sculpture with many rounded, bifurcating radial ribs; spiral sculpture of irregular commarginal growth lines. Lower margin concave, giving hooked appearance; posterior ⅔ of margin crenulate. Ligament plate long and narrow; 2 to 4 narrow, long teeth below umbo. **Habitat:** Attached to hard substrate in brackish water of inlets and bays. **Remarks:** This species typically lives on oyster reefs and other bivalve shells during times of drought.

Lioberus castaneus (Say, 1822)   Chestnut Mussel

Distribution: Florida, Texas; Caribbean, Brazil.
Size: 18 mm (¾ in).

Description: Color bluish-gray; interior grayish-white. Shape broadly oblong. Sculpture relatively smooth with commarginal growth lines covered with a brown periostracum. Posterior hinge line long and narrow; posterior adductor muscle scar large; anterior adductor muscle scar close to dorsal margin and small; margins thin. **Habitat:** Bay margins; typically associated with *Thalassia testudinium.* Found at depths from 0 to 46 m (150 ft). **Remarks:** This species typically uses its byssus to anchor itself to dead shells. Sand grains will adhere to the periostracum, which is thick and matted.

Lithophaga aristata (Dillwyn, 1817)   Scissor Date Mussel

Distribution: North Carolina to Texas; Caribbean.
Size: 12 to 50 mm (½ to 2 in).

Description: Color tannish-yellow to light brown; interior iridescent purplish-brown. Shape oblong, narrow with a projection at extreme posterior end. Sculpture rough with commarginal growth lines; shell covered with calcareous deposits. Ligament long, extends from umbo to curvature at dorsal margin; anterior adductor muscle scar long, narrow, and close to margin; adductor muscle scars not impressed. **Habitat:** Typically bores into calcareous rock. In Texas found at Flower Garden Banks. Depth range from 0 to 165 m (540 ft). **Remarks:** Recognized by the crossing of the acute, extreme posterior margins of the valves.

Lithophaga bisulcata (d'Orbigny, 1853)   **Mahogany Date Mussel**

Distribution: North Carolina to Texas-Louisiana; Bermuda; Brazil.

Size: 25 to 43 mm (1 to 1⁷⁄₁₀ in).

Description: Color brownish-mahogany; inner surface purplish-gray. Shape elongate-oval. Sculpture of commarginal growth lines; posterior portion of valve separated from frontal slope by distinct channel that extends exteriorly from umbo to ventral margin; anterior slope divided into 2 sections by second channel that extends marginally from umbo to ventral margin; shell covered with grayish-brown encrustation. **Habitat:** Bores into oysters, rocks, and coral. All along Texas-Louisiana coast at depths from 0 to 417 m (1370 ft).

Modiolus americanus (Leach, 1815)   **American Horse Mussel**

Distribution: South Carolina to Florida, Texas to Brazil; Bermuda.

Size: Adults 51 to 102 mm (2 to 4 in).

Description: Color exquisite reddish-brown to pinkish-red; inner surface glossy whitish-gray with reddish-pink showing through. Shape broadly oblong. Sculpture of commarginal growth lines; lower margin concave at byssal gape; posterior end rounded; depression or sulcus runs from umbo to posterior end of ventral margin; shell covered with brown periostracum. **Habitat:** Attached to sea whip coral by byssus; or sandy-shelly bottoms at depths from 1 to 11 m (3 to 36 ft). **Remarks:** Easily recognized, colorful shell.

Musculus lateralis (Say, 1822)   **Lateral Mussel**

Distribution: North Carolina to Florida, Texas; Caribbean.
Size: 2 to 9 mm (1⁄12 to 2⁄5 in).

Description: Color varies from greenish with brownish spots to pinkish; interior iridescent. Shape widely ovate. Sculpture of commarginal growth lines over entire outer surface; radial sculpture at both ends; no radials in middle area of shell. No cardinal teeth; lateral dentition on either side of umbo. Anterior margin crenulate; posterior ventral margin smooth. **Habitat:** Primarily associated with tunicates (_Molgula occidentalis_) and algae. Depth range from 1 to 109 m (3 to 360 ft). **Remarks:** Common on entire Texas coast. Found in abundance dredged in muddy substrate off Galveston at depths from 2 to 54 m (6 to 180 ft).

Perna perna (Linnaeus, 1785)

Brown Mussel
Distribution: Equatorial; introduced into Texas, spread to Veracruz, Mexico.
Size: 90 to 170 mm (3½ to 6½ in).

Description: Color brown externally with a purple nacreous layer internally. Shape elongate-oval. Sculpture smooth, with commarginal growth lines; ventral margin straight. Large sensory papillae on mantle margins. **Habitat:** Attached to hard substrate in high-energy zones. **Remarks:** Invasive mussel; likely transported to the region by ballast water of transatlantic vessels in 1990. Within 4 years, the edible brown mussel spread north to the jetties at Matagorda and south to Veracruz, a distance of over 1300 km (807 mi). By 1996, populations had waned; by 1998, they were difficult to find except in preferred high-energy

Pteriidae

Pteriidae is a small family of bivalves of about 20 species commonly known as winged or pearl oysters. The pteriids include pearl oysters, winged oysters, and wing shells. The shape of these bivalves is from almost round to obliquely ovate with a winged extension bordering both sides of a straight hinge line. The shell can be thin to relatively thick. Economically, this family is of much importance and has been studied accordingly. The shells are subequivalve (valves almost of equal size) to inequivalve (valves of different sizes), usually lying on the right valve, which is not as concave as the left valve. Beaks are positioned anteriorly. Sculpture is from smooth to lamellate. The periostracum is thick and dark on fresh specimens. The nacreous layer of these bivalves produces "mother-of-pearl," and although other bivalves and gastropods are capable of forming pearls, the genus _Pinctada_ excels in this respect. Pyramidellids are known to parasitize pearl oysters, and fish are known to use the shells for cover during daylight hours. In Texas Pteriidae is represented by 2 genera and 2 species whose size ranges from 38 to 76 mm (1½ to 3 in); both species are discussed here.

Pinctada imbricata (Röding, 1798) **Atlantic Pearl Oyster**

Distribution: South Carolina to Florida, Texas; Caribbean; Brazil; Bermuda.
Size: 38 to 76 mm (1½ to 3 in).

Description: Color variable from yellowish-white to greenish with radiating rays of brown to dark purple and almost black bands with white streaks; interior pearly. Shape almost round with a straight hinge area. Sculpture of commarginal, scalelike growths with extensions of radially aligned horizontal spines that extend beyond the ventral margin; hinge relatively long with a triangular posterior wing. **Habitat:** Offshore, attached to hard surfaces. Depth range from 0 to 23 m (75 ft). **Remarks:** In Texas typically found dead in beach drift.

Pteria colymbus (Röding, 1798) **Atlantic Wing Oyster**

Distribution: North Carolina to Florida, Texas; Caribbean; Brazil; Bermuda.
Size: 38 to 76 mm (1½ to 3 in).

Description: Color variable yellowish to pale brown to purplish-brown, usually with lighter, broken, radial streaks; interior pearly, inner margins not pearly. Shape obliquely oval. Sculpture relatively smooth with obscure commarginal growth lines. Posterior portion of hinge exaggeratedly extended; anterior end with an abrupt triangular bulge.
Habitat: Attached by byssus to gorgonian-type corals. Relatively common bivalve on entire Texas coast at depths from below tide line to 150 m (500 ft).

Isognomonidae

Isognomonidae is a small family of bivalves of about 20 species, commonly known as flat or tree oysters, found worldwide in tropical waters. They are usually found intertidally, but some are subtidal. A specialized ligament found in members of this family gives rise to the common name of toothed oysters. This ligament type is one of the identifying characters that separates the mallaeids from isognomonids. Shell sculpture is from smooth with undulations to fine to medium radial striations. Valves of this family are from subequal to distinctly unequal. The shape of the shell varies from circular, ovate, oblong to elongate, and are all compressed. Beaks are located anterior to the hinge line. The interior of the shell is smooth and pearly. The pallial line is typically broken into pits. The hinge plate of adults lacks teeth but possesses short, ligamental, perpendicular grooves. There are no eyespots present. Isognomonids use a byssus to attach to the underside of hard substrates and mangrove roots. In Texas Isognomonidae is represented by 1 genus and 3 species whose size ranges from 18 to 76 mm (¾ to 3 in); 2 species are discussed here.

Isognomon alatus (Gmelin, 1791)   **Flat Tree Oyster**

Distribution: Florida to Texas; Caribbean; Bermuda; Brazil.
Size: 51 to 76 mm (2 to 3 in).

Description: Color grayish-white with splotches of reddish-purple, rayed with white; interior pearly bluish-black with white, margins dull, not pearly. Shape of shell flattened, irregular fan-like, almost round. Sculpture relatively thin shell with smooth, scaly commarginal growth lines. Hinge straight with oval grooves occupied by brownish ligamental structures. **Habitat:** Nestled in crevices on jetties and other hard surfaces. Only collected alive on South Padre Island at depths from surface to 2 m (7 ft). **Remarks:** Flattest and rarest _Isognomon_ species on Texas coast.

Isognomon bicolor (C. B. Adams, 1845) **Two-toned Tree Oyster**

Distribution: Florida, Texas; Bermuda; Caribbean.

Size: 25 to 38 mm (1 to 1½ in).

Description: Color variable, typically yellowish and splotched with dark purplish-red. Shape irregularly elongate-oval. Sculpture with rough, flaky commarginal growth plates. Hinge straight with few socketlike teeth. Internally raised ledge separates the marginal region from the middle area. **Habitat:** Abundant in inlet areas from Port Aransas south. Depth range from 0 to 6 m (20 ft). **Remarks:** Middle region of inner shell is colored differently in each valve, hence the name *I. bicolor.* Panamic Vicariant (i.e., species separated by a barrier, in this case, the Isthmus of Panama): *I. recognitus* (Mabille, 1895).

Ostreidae

Ostreidae is a ubiquitous family of bivalves found worldwide except at the poles. Although quite abundant, there are only about 50 species in the family. Shell size is from medium to large (25 to 330 mm, or 1 to 13 in). Shell structure is irregular and circular, narrowly ovate, and usually compressed. The left, or attached, valve is the bottom valve and is typically flatter than the right, or upper, valve. Sculpture is smooth to coarse and may have growth lines or radial ribs that can be strong or weak, with spines occurring on some species. Shell margins are generally strong and folded or scalloped. The interior is glassy white except for the muscle scars, which are tinged green to purplish-brown. Some ostreids have small ridges, pits, or beads lined up on both sides of the hinge. Ostreid larvae survive by attaching themselves to a hard surface. Ostreids begin life as males and end as females. In Texas waters Ostreidae is represented by 4 genera and 5 species whose size ranges from 12 to 152 mm (½ to 6 in); 2 species are discussed here.

Crassostrea virginica (Gmelin, 1791) **Eastern Oyster**

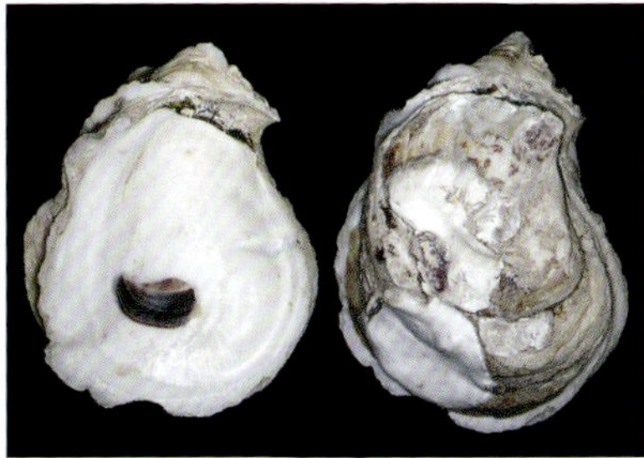

Distribution: Gulf of St. Lawrence; Gulf of Mexico, Texas; Caribbean.
Size: 51 to 152 mm (2 to 6 in).

Description: Color grayish to yellowish-white with bands of reddish-purple; interior white with purple muscle scar in posterior margin. Shape semicircular to elongate. Sculpture semi-smooth with irregular scales and growth ridges; ventral end sometimes with platelike growths. **Habitat:** On hard and soft bottoms at depths from 0 to 79 m (260 ft). **Remarks:** Highly prized commercial species. Large numbers grow together to form intertidal and subtidal oyster reefs within Texas bays. **Synonyms:** *C. brasiliana* Lamarck, 1819; *C. floridensis* (Sowerby, 1841).

Ostrea equestris (Say, 1834) **Crested Oyster**

Distribution: Virginia to Texas; Caribbean, Brazil.
Size: 25 to 76 mm (1 to 3 in).

Description: Color whitish, covered with brownish-purple; interior whitish, splotched with yellowish-green. Shape triangular to subcircular. Sculpture variable, typically compressed and smooth, scaly, or ridged. Margin smooth or angularly serrate. Upper hinge with small ridges that fit into bottom hinge pits; bottom valve typically deep with margin elevated. **Habitat:** On hard surfaces at depths from shoreline to 100 m (330 ft). **Remarks:** One of the most common ostreids taken alive on Texas coast. Found at Seven and One-half Fathom Reef and in the hypersaline lagoons of Texas. **Synonym:** *Ostreola equestris* (Say, 1834).

Pinnidae

Pinnidae is a small family of bivalves comprising about 20 species, commonly known as pen shells, wing shells, or fan shells, found worldwide. They are typically located in tropical waters, but some inhabit temperate climes. These bivalves are usually large animals from 100 to 790 mm (4 to 31 in) and are triangular and thin. Outer shell sculpture is of radial ribs that typically appear posteriorly and may be spinose. Valves are usually equal and may be flexible when fresh. The shells become brittle and difficult to handle when dried and worn. The hinge line is long and almost straight. Beaks are located toward the anterior end of the hinge margin. Adult shells lack teeth. There is no pallial line. Adductor muscles allow the shell to contract when disturbed. These creatures are typically buried in mud and/or shell fragments, where they attach byssally with their rounded posterior end exposed and slightly agape. In Texas Pinnidae is represented by 2 genera and 5 species whose size ranges from 100 to 305 mm (4 to 12 in); 2 species are discussed here.

Atrina seminuda (Lamarck, 1819) — **Half-naked Pen Shell**

Distribution: North Carolina to Texas to Argentina.
Size: 127 to 229 mm (5 to 9 in).

Description: Color translucent bluish-brown, sometimes splotched with purplish-brown. Shell fan shaped. Sculpture of coarse radiating ribs becoming broader toward the ventral margin, with many tubular spines projecting from the ribs. Interior muscle scar restricted to the border of the nacreous layer.
Habitat: Inlet-influenced bays and offshore. Found alive at depths from low-tide line to 24 m (80 ft). **Remarks:** *A. seminuda* is sometimes found on Texas beaches washed ashore alive in great numbers, usually after storms when they are "uprooted" from the bottom. Fragments of this shell are frequently dredged from depths of 256 m (840 ft).
Synonyms: *A. alta* (Sowerby I, 1835); *A. listeri* (d'Orbigny, 1846).

Atrina serrata (Sowerby I, 1825)   **Saw-toothed Pen Shell**

Distribution: North Carolina to Florida, Texas; Caribbean.
Size: 152 to 305 mm (6 to 12 in).

Description: Color translucent yellowish-gray to greenish. Shell fan shaped. Sculpture of somewhat thin radial ribs becoming finer on ventral slope, with relatively small modular projections. Hinge line almost straight on upper portion, concave on lower part of margin. Nacreous layer covers ¾ of interior of shell; posterior adductor muscle scar round, positioned deep inside nacreous layer; anterior adductor muscle scar small. **Habitat:** Inlet-influenced areas of bays on Texas coast. Depth range from 0 to 42 m (138 ft). **Remarks:** Sometimes taken alive when washed ashore on Texas beaches.

Limidae

Limidae is a moderate-sized family of bivalves consisting of about 125 species commonly known as file clams. The Limidae are found throughout tropical and temperate waters. Biological attributes such as a colorful mantle, pallial tentacles, and swimming style make these mollusks a commonly collected and illustrated family. Shells can be strong or fragile, ovate, or circular to triangular. These bivalves have small "ears," or extensions, on both sides of the hinge line. Sculpture can be smooth or with spinose or scaly radial ribs. The ligamental area is wide, triangular, and located above the hinge line. The outer mantle fold is equipped with eyespots or photoreceptors. The shells are equivalve. Dentition is taxodont and weak. Limids inhabit coral reefs, attached by a byssus, or are free swimming and hide under hard substratum. Limids are able to swim by contracting their adductor muscles strongly and flapping their valves. In Texas, Limidae is represented by 5 genera and 7 species whose size ranges from 5 to 76 mm (⅕ to 3 in); only the most common Texas species is discussed here.

Limaria pellucida (C. B. Adams, 1846)   **Antillean File Clam**

Distribution: North Carolina to Florida, Texas to Brazil; Bermuda.

Size: 10 mm (⅖ in).

Description: Color translucent yellowish-white. Shape semi-ovate, shell somewhat broad. Sculpture thin with thread-like semi-separated radial ribs crossed by commarginal growth lines. Hinge almost straight; ventral margin crenulate by ribs extending onto ventral margin in fresh shells. **Habitat:** In Texas offshore in sandy calcareous environments at depths from 22 to 54 m (72 to 186 ft). Depth range from 0 to 183 (600 ft). **Remarks:** Some confusion exists between this species and another similar, larger species, giving rise to the "*pellucida*" species complex.

Pectinidae

Pectinidae is a large family of bivalves of approximately 350 species. They are found worldwide in nearly all depths from shallow, sub-tidal waters to the abyss. Climatic range is also extensive, from tropical regions to the poles. Commonly known as scallops, these bivalves are important commercially as food. Because they come in many different colors, shapes, and sizes, they are also important to shell collectors. Shell size is from small (6 mm, or ¼ in) to large (300 mm, or 12 in). Pectinid shells typically get narrow at the beaks and possess flattened triangular projections called "ears." Eyes sensitive to light are found along the mantle edge. Sculpture is of radial and concentric ribs; however, some pectinids have smooth shells. Most scallops can "swim" but do so only when aroused. Those attached by byssal threads can escape by clapping their valves. In Texas, Pectinidae is represented by 6 genera and 8 species whose size ranges from 18 to 152 mm (¾ to 6 in); 2 species are discussed here.

Argopecten irradians amplicostatus (Dall, 1898)　**Western Bay Scallop**

Distribution: Texas to Mexico; Colombia.
Size: 38 to 102 mm (1½ to 4 in).

Description: Color of exterior of upper valve grayish-brown with irregular, concentric, darker markings; interior white; bottom valve uniformly white. Shape subcircular. Sculpture of thick, modular ribs; exterior of auricles with radial ribs and concentric threads, interior of auricles smoothish with crenulations at margins; ventral margins serrate. **Habitat:** On muddy sand; found alive in bays of Texas. Depth range from 0 to 26 m (85 ft). **Remarks:** _A. i. amplicostatus_ (Dall, 1889), western bay scallop; _A. i. concentricus_ (Say, 1822), southern bay scallop; and _A. i. irradians_ (Lamarck, 1819), northern bay scallop, may be clines rather than subspecies.

Nodipecten fragosus (Linnaeus, 1758)   **Northern Lion's Paw**

Distribution: North Carolina to Florida, Texas; Caribbean; Brazil.

Size: 76 to 152 mm (3 to 6 in).

Description: Color typically uniformly dark maroon to red, may be orange and sometimes yellow; interior pearly toward dorsal center, maroon-red on margins. Shape fanlike. Sculpture of distinct, rough, large, nodular ribs with finer ribs on and in between large ones crossed by concentric, shelflike ridges giving nodulose, knucklelike appearance; interior smooth with exterior nodulose sculpture coming through. **Habitat:** Common on offshore banks and coral reefs. Depth range from 16 to 185 m (95 to 610 ft). **Remarks:** Valves have been found from Matagorda to South Padre Island at depths from 16 to 100 m (54 to 330 ft). Formerly known as _Lyropecten nodosus_ (Linnaeus, 1758), which is a Caribbean and South American species.

Spondylidae

Spondylidae is a small family of bivalves of 1 genus (_Spondylus_) and about 40 species. These bivalves are commonly known as thorny oysters and are found in all tropical marine habitats. They are distinguished by their relatively large size and presence of long spines. Shell shape is circular to ovate with convex valves. The right valve is usually cemented to a hard substrate and is more flattened than the left valve. The beaked area and ligament are higher in the right valve than the left. "Ear" extensions are small to medium. The exterior part of the left valve has irregular, spiny radial ribs of different shapes and sizes. Both valves are equipped with sturdy, interlocking teeth and sockets that are important in keeping the valves aligned. The interlocking valves, spines, and relatively heavy cemented shells of spondylids are useful for protection against predators. Like the pectinids, the spondylids have light-sensitive eyes found in short intervals along the mantle edge. In Texas, Spondylidae is represented by 2 species whose size ranges from 25 to 230 mm (1 to 9 in); only the most common species is discussed here.

Spondylus americanus (Hermann, 1781)   **Atlantic Thorny Oyster**

Distribution: North Carolina to Florida, Texas; Caribbean to Brazil.

Size: 76 to 230 mm (3 to 9 in).

Description: Color variable, base whitish; umbones and auricles yellow, orange, red, purple to reddish-brown; inner portion of shell smooth and white. Shape wide and ovate to almost spherical. Sculpture of shell thick, with numerous long and short spiny projections on exterior of uncemented top valve. Bottom valve attached to hard substrate typically flattencd. Hinge of interlocking teeth (see inset). **Habitat:** Cemented to offshore reefs, banks, and oil and gas platforms at depths from 10 to 168 m (33 to 550 ft). **Remarks:** Specimens from Texas grow large, to over 229 mm (9 in).

Plicatulidae

Plicatulidae is a small family of bivalves of 1 genus (*Plicatula*) and approximately 10 species. They are typically found in shallow, tropical waters. The shells are from oval to triangular, becoming narrower toward the beaks with crenulate margins. Valves are plicate and do not develop "ears" on the sides of the hinge line. Sunken sturdy hinge plates on both valves possess secondary teeth and sockets that interlock to keep the valves in place. The interior of the valve has a small, somewhat rounded muscle scar. Typically, the shells cement their right valve to hard substrate. In Texas, Plicatulidae is represented by 1 species.

Plicatula gibbosa (Lamarck, 1801)

Atlantic Kitten's Paw

Distribution: North Carolina to Florida, Texas; Caribbean; Bermuda; Brazil.
Size: 25 mm (1 in).

Description: Color yellowish-white; interior white. Shape triangular. Sculpture strong, thick, and relatively smooth with commarginal growth ridges and well-developed, thick, pipelike radial ribs that become broader toward ventral margin. Subcircular muscle scar toward flared margin. Umbones thick and narrowed. Margins smoothish and broadly crenulate. **Habitat:** All along the Texas coast at depths from shoreline to 110 m (360 ft). **Remarks:** Shell design and habitat are so diverse that some authors believe 2 species may be involved. However, similarities between all specimens make it difficult to separate them.

Anomiidae

Anomiidae is a small family of bivalves of approximately 25 species, commonly known as jingle shells, found throughout the world. Shell outline is irregular, usually distorted, but somewhat circular to ovate. Shell sculpture is typically with rough radial riblets and usually mirrors the contour of the substrate. Shells are usually fragile and semi-translucent when fresh, with the right valve thinnest. In most cases the upper left valve is convex; the lower right valve is flat with a hole present at the upper margin where the byssus passes. Valves are unequal. The posterior (almost central) muscle is small, and the anterior muscle has been lost. The hinge is without teeth, and valves are joined by an interior ligament in the right valve. Attachment scars produced by pedal and byssal retractor muscles are used to distinguish the species. In Texas Anomiidae is represented by 2 genera and 2 species whose size ranges from 25 to 51 mm (1 to 2 in); 1 species is discussed here.

Anomia simplex (d'Orbigny, 1853)   **Common Jingle**

Distribution: Massachusetts to Florida, Texas; Brazil; Bermuda.

Size: 25 to 51 mm (1 to 2 in).

Description: Color translucent yellow to pale orange, typically with a silvery gloss. Shape subcircular; shell flattened, thin but sturdy. Sculpture smooth with commarginal growth ridges toward exterior ventral margins, with a hole located on lower valve where the byssus passes through. Interior smooth; muscle scar present in triangular area, which extends from the beak to middle interior part of shell. **Habitat:** Attached to hard substrates; from shallow bays to offshore at depths from 0 to 128 m (420 ft). **Remarks:** One of the most common species on the Texas coast.

Crassatellidae

Crassatellidae is a relatively small family of bivalves of approximately 30 species. They are found in shallow tropical and warm temperate waters. Shell shape is from triangular to subquadrate with the posterior end drawn out and angled. The shell is typically robust, ovate, and heavy. Sculpture may be smooth or finely ridged. Muscle scars are ovate and almost equal in size. Attachment scars are located interiorly toward the dorsal margin. Located in the beaked area next to the hinge is a well-defined scar. The zone of insertion created by the strong pallial muscles is located ventrally and clearly marked by the pallial line. In the hinge area are 2 cardinal teeth in the left valve and 1 to 3 cardinal teeth in the right valve. In Texas, Crassatellidae is represented by 2 genera and 3 species whose size ranges from 1.5 to 64 mm (1/16 to 2½ in); 1 species is discussed here.

Crassinella lunulata (Conrad, 1834) Lunate Crassinella

Distribution: Massachusetts to Florida to Texas; Brazil; Bermuda.

Size: 6 to 8 mm (¼ to ⅓ in).

Description: Color white, tinged with pink; interior glossy brown. Shape triangular and compressed. Sculpture smoothish with concentric growth lines; radial sculpture of microscopic threads. Umbones acute; dorsal margin almost straight; ventral margin smooth and rounded; 2 cardinal teeth in each valve; long ligamental depression in both valves under umbo. **Habitat:** Common in bays and offshore shallow waters; in sand or mud at depths from shoreline to 128 m (420 ft). **Remarks:** This species differs from other *Crassinella* by its size; no other *Crassinella* grows larger than 5 mm (⅕ in).

Carditidae

Carditidae is a large family of bivalves of approximately 200 species found worldwide. Shell shape is from elongate to almost round, wide and ovate, or subtriangular. Sculpture is typically with strong ribs. The ribs can be intersected by commarginal elements that form scales. The shell itself is thick and strong. The beaks are arched forward and equipped with an external ligament. The margins are either scalloped or toothed. The beak of carditids is turned forward. The siphon is lacking in the carditids. The periostracum is usually dense and dark. In Texas, Carditidae is represented by 1 species, whose size ranges from 25 to 38 mm (1 to 1½ in).

Carditamera floridana (Conrad, 1838) — Broad-ribbed Carditid

Distribution: Florida, Texas; Mexico.

Size: 25 to 38 mm (1 to 1½ in).

Description: Color whitish to grayish with reddish-brown maculations arranged concentrically on outer surface of shell; inner surface white. Shape oblong-quadrate. Sculpture of flattened, beaded, distinct radial ribs with concentric threads between radials; ventral margin below hinge area scalloped. Hinge distinct; large cardinal tooth in right valve, smaller one in left valve. Sturdy shell. **Habitat:** Typically collected in hypersaline bays and beaches. Depth range from 0 to 190 m (623 ft). **Remarks:** Shells are commonly used in shell craft. Found southward from Matagorda Island to South Padre Island.

Pandoridae

Pandoridae is a small family of bivalves of about 25 species. They are found in all oceans, but most are located in cooler waters of the northern hemisphere. Shell outline is crescent shaped and compressed. Sculpture is typically smooth, but the left valve may have commarginal growth lines. The front part of the shell is shorter and more circular than the longer hind end. The right valve is more convex than the more compressed left valve, which overlaps at the top margin, and has one or more ridges extending from the beak to the postero-ventral margin of the shell. The pallial line is interrupted and plain, and the pallial sinus is absent. Muscle scars are almost equal. Poorly developed spoon-shaped element (chondrophore) houses an internal ligament for attachment. The hinge margin is straight, with a sunken beak, and the hinge area has 1 to 3 small, toothlike ridges found close to the margin. There is no true dentition. These bivalves typically burrow below the surface in mud, sand, or gravel with siphons projecting above the surface. In Texas, Pandoridae is represented by 1 genus and 3 species whose size ranges from 8 to 25 mm (1/ to 1 in); 1 species is discussed here.

Pandora trilineata (Say, 1822)   **Threeline Pandora**

Distribution: North Carolina to Florida to Texas.

Size: 18 to 25 mm (¾ to 1 in).

Description: Color creamish-white; inner surface pearly. Shape half moon, squared off at anterior margin, somewhat pointed posteriorly, distinctly compressed. Sculpture smoothish with irregular, wavy commarginal growth lines; left valve with 2 radial ribs and a sulcus that extends from umbo to ventral angle; hinge area with a resilium (internal ligament) and lithodesma (calcareous strengthening of the internal ligament). **Habitat:** Bay and lagoon centers in soft clay or mud. Rarely collected offshore; typically collected alive at depths from 2 to 4 m (6 to 12 ft). Depth range from 0 to 91 m (300 ft). **Remarks:** A bay dweller.

Lyonsiidae

Lyonsiidae is a small family of bivalves made up of approximately 20 species with almost a worldwide distribution, but it is not recorded for Australia. Shell outline is oval, oblong, sometimes almost circular, and frequently irregularly shaped. Sculpture can be smooth, with commarginal growth lines or radial ribs. Valves are subequal and overlap ventrally. The hinge is irregular and without dentition. Beaks are situated centrally to the longitudinal axis of the shell. The posterior end has a siphonal gape and is oval and wider than the closed, rounded anterior end. The posterior muscle scar is small and almost equal in size to the anterior muscle scar. The fused margins possess a large to small pedal aperture with a fourth pallial aperture. The pallial sinus is relatively small, and the pallial line is distinct. Shells include a high content of nacre, which allows them to be thin yet sturdy. Siphonal photoreceptors allow the lyonsiids to detect predators in the water column. These bivalves typically nestle in crevices or gravel or are associated with compound ascidians. In Texas, Lyonsiidae is represented by 2 genera and 3 species whose size ranges from 12 to 25 mm (½ to 1 in); 1 species is discussed here.

Lyonsia hyalina (Conrad, 1831)   **Glassy Lyonsia**

Distribution: Florida to Texas.

Size: 18 mm (¾ in).

Description: Color white, somewhat translucent. Shape elongate-oval. Sculpture of fine radial lines crossed with obscure commarginal threads; anterior end smoothly rounded; posterior end tapered and truncate with a ridge from umbo to postero-ventral angle. Hinge area edentate with a long ossicle below the beaks for attachment. **Habitat:** Offshore in sandy bottoms. In Texas found alive at depths from 22 to 51 m (72 to 168 ft). Depth range from 0 to 62 m (204 ft). **Remarks:** Difficult to distinguish from *L. floridana,* but ventral margin not as rounded and posterior dorsal margin straighter.

Periplomatidae

Periplomatidae is a small family of bivalves of about 28 species. These bivalves are usually found in cold and temperate waters of the northern hemisphere with some species found in Australia and New Zealand. Shell outline is broad and ovate to almost oblong. Shell sculpture is usually of indistinct commarginal ridges, rarely with radial sculpture. The right valve is typically larger and more convex than the left valve. The hinge area is without dentition. The posterior muscle scar is almost round; the anterior muscle scar is slender, elongate, and placed close to the shell margin. The anterior margin is circular; the posterior margin is beak shaped and short. Both valves have a lower beak crack directly below the beaks that runs downward to about the same distance as the spoon-shaped element. The pallial line is distinct, and the pallial sinus is deep. Mantle lobes are fused except for the siphonal gape, pedal gape, and fourth pallial apertural gape. In Texas, Periplomatidae is represented by 1 genus and 2 species whose size ranges from 8 to 12 mm (⅓ to ½ in); 1 species is discussed here.

Periploma margaritaceum (Lamarck, 1801)   **Unequal Spoon Clam**

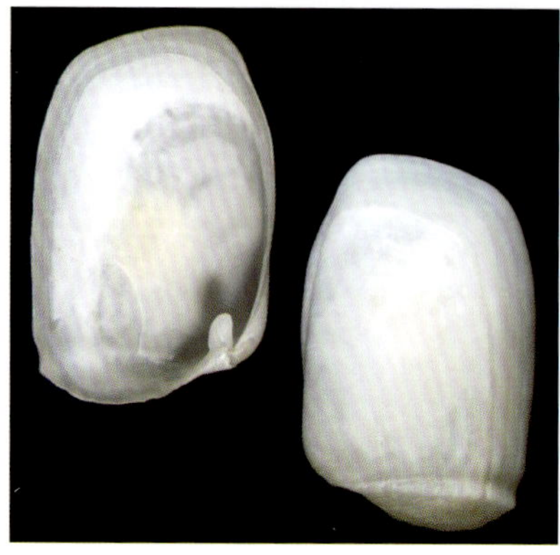

Distribution: South Carolina to Florida to Texas.

Size: 18 to 25 mm (¾ to 1 in).

Description: Color opaque white; inner surface glassy white. Shape rectangularly ovate. Sculpture smoothish with faint commarginal growth lines and microscopic beads, beads most pronounced posteriorly; anterior end longer than posterior end, well rounded; posterior end abrupt with almost vertical ridge running from umbo to ventral angle. Hinge area with spoon-type ligamental shelf pointed anteriorly; accessory plate above. **Habitat:** Found in sand or mud at bay margins at depths from shoreline to 14 m (45 ft). **Remarks:** Shells regularly found in beach drift, often in pairs in bays, and as single valves along Gulf beaches in Texas.

Lucinidae

Lucinidae is a large family of bivalves comprising approximately 200 species found throughout the world. Almost all lucinids live in mud or soft sand in areas of low nutrients and high sulfide content. Most lucinids live in symbiosis with a bacterium that is able to fix carbon dioxide and to contribute to sulfide oxidation in its depleted oxygen environment. Shell shape is usually subcircular; valves are equal. The beaks are small, sturdy, centrally located, and turned forward. Sculpture may be smooth, radial, concentric, or both radial and concentric, forming a cancellate appearance. Lucinids typically have 1 to 2 radial furrows or ridges on the external posterior region of the valve. The hinge plate is weak, typically with 2 cardinal teeth in each valve found beneath the beaks. Lateral dentition is variable and may be strong, weak, or absent. In Texas, Lucinidae is represented by 10 genera and 11 species whose size ranges from 6 to 102 mm (¼ to 4 in); 7 species are discussed here.

Anodontia alba (Link, 1807) — Buttercup Lucine

Distribution: North Carolina, Texas, Gulf of Mexico; Caribbean; Bermuda.

Size: 38 to 64 mm (1½ to 2½ in).

Description: Color whitish; interior whitish stained with yellow-orange. Shape subcircular, thin. Sculpture smoothish with commarginal growth lines and irregular wrinkles. Beaks small, not touching; depressed posterior sulcus; lunule small, ovate, part from right valve overlaps left. Margin smooth. Ligament on embedded shelf. **Habitat:** In bays on sandy bottoms and shallow, offshore waters at depths from shoreline to 22 m (72 ft). **Remarks:** Relatively flat shell.

Anodontia schrammi (Crosse, 1876) — Chalky Buttercup Lucine

Distribution: North Carolina, Florida, Texas; Cuba; Bermuda.

Size: 51 to 102 mm (2 to 4 in).

Description: Color white; interior chalky white. Shape subcircular, globose. Sculpture with granular surface and numerous concentric, moderately raised growth lines; growth lines more numerous on area of sulcus. Beaks small, located anterior to midpoint of shell; lunule small and narrow, converges at midline without overlap. **Habitat:** Common in dredged material at depths from 0 to 128 m (420 ft). **Remarks:** Britton's (1970) dissertation on the family Lucinidae corrects this species name from *A. philippiana* (Reeve, 1850). **Synonym:** *Lucina schrammi* (Crosse, 1876).

Codakia orbicularis (Linnaeus, 1758) **Tiger Lucine**

Distribution: Florida, Texas, Gulf of Mexico; Caribbean, Brazil; Bermuda.
Size: 64 to 89 mm (2½ to 3½ in).

Description: Color uniformly white, stained yellow or pink. Shape subcircular. Sculpture of fine radial ribs crossed by fine concentric ribs, giving reticulate appearance. Beaks touching; lunule small, heart shaped, filled by platelike projection in right valve, excavated in left valve. Hinge with 2 cardinal teeth and 1 anterior lateral tooth in each valve. **Habitat:** In Texas, found on shell bottoms and beaches at depths from shoreline to 93 m (305 ft). **Remarks:** In Texas from Port Aransas southward.

Parvilucina crenella (Dall, 1901) **Many-lined Lucine**

Distribution: North Carolina to Florida, Texas; Brazil.
Size: 6 to 9 mm (¼ to 3/8 in).

Description: Color uniformly white. Shape subcircular. Sculpture of flattened, concentric threads and radial striae, creating reticulate appearance. Sulcus extending from beak to hind end of ventral margin. Hinge area narrow; cardinal area distinct; laterals separated from cardinals by narrow groove. Anterior adductor muscle scar elongate, posterior adductor muscle scar subovate; pallial line irregular. Ventral margin crenulate. **Habitat:** In beach drift on Texas coast to a depth of about 91 m (300 ft). **Synonyms:** *P. multilineata* (Toumey and Holmes, 1857); *Lucina crenulata* (Dall, 1901).

Phacoides pectinata (Gmelin, 1791)   **Thick Lucine**

Distribution: North Carolina to Florida, Texas to Brazil.
Size: 25 to 64 mm (1 to 2½ in).

Description: Color whitish, sometimes stained with orange; inner shell yellowish with tinged orangish margin. Shape subcircular and flattened, equivalve. Sculpture of erect, irregular, concentric, scaly ribs with finer concentric threads between. Hinge area long; cardinal teeth feeble, lateral teeth strong; lunule abrupt and slender. Anterior adductor muscle scar long, narrow, and impressed; posterior adductor muscle scar ovate; pallial line irregular, narrow, and somewhat weak. **Habitat:** All coastal bays of Texas and offshore to a depth of 40 m (130 ft). **Synonyms:** *Lucina jamaicensis* (Chemnitz, 1784); *L. pectinata* (Gmelin, 1791).

Radiolucina amianta (Dall, 1901)   **Miniature Lucine**

Distribution: North Carolina, Florida, Texas; Caribbean to Brazil.
Size: 6 to 9 mm (¼ to ⅖ in).

Description: Color brownish-white; inner surface chalky white. Shape subcircular to somewhat triangular, equivalve. Sculpture of distinctly raised, wide radial ribs crossed by concentric ribs that appear flat and scalelike where they intersect. Lunule medium in size; anterior sulcus distinct; anterior cardinal obscure, posterior cardinal subtriangular, fused with dorsal portion of lunule; 1 lateral tooth at either end of hinge area. **Habitat:** Bays on sandy bottoms to offshore Texas coast. Depth range is from below tide line to 1170 m (3838 ft).

Stewartia floridana (Conrad, 1833)   **Florida Lucine**

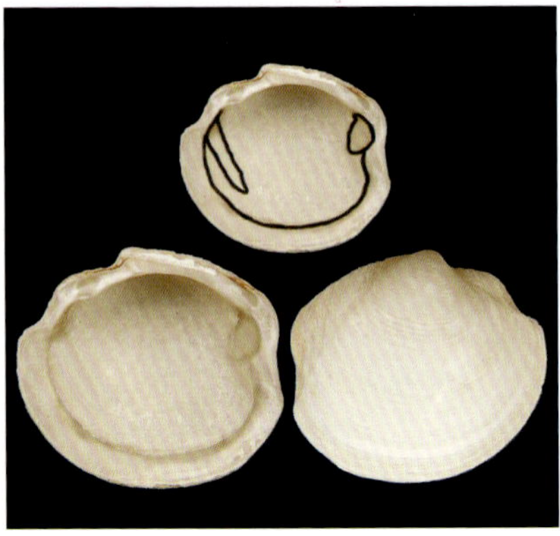

Distribution: Florida to Texas.
Size: 38 mm (1½ in).

Description: Color tannish; interior off white; periostracum deciduous and yellow. Shape subcircular. Sculpture of concentric, elevated growth lines that get wider toward the ventral margin. Lunule small, weak, long, with a slightly excavated area under the umbones; hinge area thickened; ligament long; cardinal region with 2 weak cardinal teeth. Pallial line simple. **Habitat:** In sand or mud bottoms of coastal bays at depths from shoreline to 46 m (150 ft). **Remarks:** Some authors placed this species in the genus _Pseudomiltha,_ and others, in the genus _Lucina._ We place it in the genus _Stewartia,_ following the recent work of Mikkelsen and Bieler (2008). **Synonyms:** _L. floridana_ (Conrad, 1833); _P. floridana_ (Conrad, 1833).

Ungulinidae

Members of the Ungulinidae were formerly in the family Diplontidae. This is a small-sized family of bivalves comprising about 50 species. These organisms are found throughout the world in temperate and tropical waters. Their thin shells vary from subcircular to ovate and are usually oblique and inflated. The beaks are turned backward and may be distinct or low. Sculpture is usually smooth or with fine concentric ribs, rarely with pits or beads. The valves are equal. The anterior and posterior muscle scars are elongate. Beneath the beaks on the hinge are 2 cardinal teeth, found on both valves; lateral teeth are absent. The pallial line is entire; no pallial sinus is present. Diplodons are typically nestlers and live in rocks, crevices, empty mollusk shells, or in sand or gravel. In Texas, Ungulinidae is represented by 3 genera and 4 species whose size ranges from 8 to 25 mm (⅓ to 1 in); 2 species are discussed here.

Phlyctiderma semiaspera (Philippi, 1836)   **Pimpled Diplodon**

Distribution: North Carolina to Florida, Texas; Caribbean; Brazil.

Size: 12 mm (½ in).

Description: Color uniformly chalky white. Shape inflated, subcircular. Sculpture smooth, similar to that of *Diplodonta punctata* (see Tunnell et al. 2010: 343) with fine, microscopic growth threads composed of microscopic punctae. Hinge line narrow; cardinal area made up of long tubule for ligamental attachment; no lateral teeth. Margin smooth. **Habitat:** In sand or mud, often under rocks and in shells. All along the Texas coast at depths from shoreline to 104 m (340 ft).

Phlyctiderma soror (C. B. Adams, 1852)   **Sister Diplodon**

Distribution: North Carolina to Texas-Louisiana; Caribbean.

Size: 18 mm (¾ in).

Description: Color uniformly chalky white. Shape subcircular. Sculpture of irregularly spaced, commarginal growth lines and fine, concentric lamellae between larger growth lines. Beaks positioned centrally, slightly pointing anteriorly; posterior hinge long and narrow, anterior hinge wider; cardinal area distinct. Anterior adductor muscle scar ovate; posterior adductor muscle scar oval and elongate. Ventral margin smooth. **Habitat:** Sandy bottoms along Texas-Louisiana coast at depths from shoreline to 128 m (420 ft). **Remarks:** Rarely dredged in coastal bays. **Synonym:** *Diplodonta soror* (C. B. Adams, 1852).

Chamidae

Chamidae, commonly known as the jewelbox family, is a relatively small family of bivalves of about 50 species found in temperate and tropical waters worldwide. These bivalves are either permanently or temporarily cemented to a hard substrate during some time in their life. Some species are attached as juveniles and free swimming as adults. Others are attached throughout their life. Shell shape is thick, irregular, and somewhat circular to oval. Strong, complex sculpture may be concentric, leafy, platelike, or combinations of these that may be accompanied by flattened, imbricate frills or axial, elongate spines. Shells are inequivalve in attached forms; however, in unattached forms there is little difference. Attachment by these shells is dependent on environmental factors and may be from the right or left valve, causing intraspecific variation that has resulted in taxonomic confusion for some species. In Texas, Chamidae is represented by 2 genera and 6 species whose size ranges from 25 to 76 mm (1 to 3 in); 2 species are discussed here.

Arcinella cornuta (Conrad, 1866)

Florida Spiny Jewelbox

Distribution: North Carolina to Florida, Texas.
Size: 25 to 51 mm (1 to 2 in).

Description: Color uniformly white or splotched purplish-pink. Shape somewhat oval, deep, thick, and heavy. Sculpture of strong radial ribs with large scaly spines and scalelike plates between radials. Margin crenulate. Lunule distinct, heart shaped; umbones curved forward and down. **Habitat:** Sandy, shelly bottoms in offshore waters at depths from 0 to 73 m (240 ft).
Remarks: *A. cornuta* differs from *A. arcinella* (Linnaeus, 1767) primarily by having fewer spiny radial ribs and being thinner and not as inflated (see Tunnell et al. 2010).

Chama congregata (Conrad, 1833)

Little Corrugate Jewelbox

Distribution: North Carolina to Texas; Brazil; Bermuda.
Size: 33 mm (1⅓ in).

Description: Color brownish-orange on grayish base; interior white with reddish-purple splotches. Shape subcircular. Sculpture of short, flat, radial corrugations on both valves. Inner margin denticulate.

Small prodissoconch (larval shell). **Habitat:** Typically attached to hard substratum offshore and on Gulf jetties. Common offshore bivalve; in Texas found alive at depths from 11 to 78 m (36 to 258 ft). Depth range from 0 to 101 m (330 ft).

Lasaeidae

Lasaeidae is a problematic family of small to minute white bivalves found within the confused superfamily Galeommatoidea. Shells are typically white, fragile, subcircular to ovate to elongate-oval, usually less than 12 mm (½ in). The anterior and posterior muscles are subequal. Retractor and protractor foot muscles are present. These bivalves live in association with many other invertebrate species. Some live beneath rocks, and others live in tunnels or tubes or commensally on the bodies of annelid worms, other bivalves, crustaceans, or holothurians. In Texas, Lasaeidae is represented by 8 genera and 8 species whose size ranges from 1 to 16 mm (⅕5 to ⅔ in); 3 species are discussed here.

Aligena texasiana (Harry, 1969) — Texas Aligena

Distribution: Texas, Louisiana.
Size: 5 mm (⅕ in).

Description: Color uniformly opaque white with a tan periostracum. Shape rounded to ovate to subtriangular to subquadrate, inflated. Sculpture smoothish with fine, irregularly spaced, commarginal growth lines. Prodissoconch circular well defined. Umbones touch; resilium abrupt, attached along a slight ridge beneath dorsal margin; large, single, spoonlike tooth in right valve, fits into left valve. **Habitat:** Mud bottoms of bays. On the Texas coast this species is commonly found dead in depths from shoreline to 16 m (52 ft). **Remarks:** Rarely collected alive.

Lepton lepidum (Say, 1826) — Graceful Lepton

Distribution: South Carolina, Texas.
Size: 5 mm (⅕ in).

Description: Color translucent white. Shape subquadrate. Sculpture smooth, fragile, with faint, microscopic, commarginal growth lines. Umbones erect and positioned toward the center of upper margin; single, small, obscure cardinal hook in left valve, absent in right; resilifer (depression on the hinge plate) internal, abrupt, and sturdy. Pallial line simple. **Habitat:** Commensal; attached to other invertebrates by byssus and has been seen attached to the underside of crustaceans. May be a surf-zone species. **Remarks:** This species is typically found in the surf zone; not found in bays or offshore dredgings.

Mysella planulata (Stimpson, 1851)   **Plate Mysella**

Distribution: Nova Scotia to Texas; Caribbean. **Size:** 3 mm (⅛ in).

Description: Color uniformly translucent opaque white. Shape broadly ovate, somewhat compressed. Sculpture smooth with irregular, commarginal growth lines. Small acute umbones located anteriorly, but pointing posteriorly. Pallial apparatus simple. **Habitat:** Shallow water; attached to flattened or hard surfaces. Found in all Texas bays in depths from shoreline to 2 m (7 ft). Depth range from 0 to 88 m (290 ft). **Remarks:** Common in beach drift.

Hiatellidae

Hiatellidae is a small family of bivalves of approximately 25 species. They are mostly found in temperate waters of the north. Shell shape is typically oblong and equivalve. The shell is generally thick, and sculpture is usually of wavy commarginal growth lines with valves narrowly to broadly gaping. The well-developed periostracum overlaps the mantle edge and siphonal sheath. Beaks are located anteriorly or in front of the middle of the upper or dorsal margin. Muscle scars are almost equal. A single, central tooth is reduced to obsolete; lateral teeth are not present. The pallial sinus is distinct, and the pallial line is discontinuous and impressed. Siphons are almost entirely fused. Hiatellids bore into soft rock, mud, or sand or nestle in crevices or old bore holes, using a byssus for attachment. In Texas, Hiatellidae is represented by 2 genera and 2 species whose size ranges from 25 to 152 mm (1 to 6 in); 2 species are discussed here.

Hiatella arctica (Linnaeus, 1767)　　　　　　　　Arctic Hiatella

Distribution: Texas; Arctic seas to deep water of Caribbean. **Size:** 25 mm (1 in).

Description: Color uniformly chalky white. Shape elongate-rectangular. Sculpture of irregular, rough commarginal growth lines; radial ribs on posterior ventral angle becoming scaly where they cross. Umbones positioned anteriorly; teeth indistinct. Pallial line interrupted; pallial sinus broad but irregular. Periostracum deciduous. **Habitat:** Hypersaline bays, inlets, and offshore banks at depths from 0 to 101 m (330 ft). **Remarks:** Boring and nestling patterns of *H. arctica* produce a variety of shapes and sculpture for this species.

Panopea bitruncata (Conrad, 1872)　　　　　　　Atlantic Geoduck

Distribution: North Carolina to Florida to Texas. **Size:** 127 to 190 mm (5 to 7½ in).

Description: Color uniformly white. Shape oblong-oval with a gape at both ends. Sculpture of irregular, wavy commarginal growth lines. Umbones low, recurved inward; hinge with 1 large central tooth with a broad plate for ligamental attachment. Anterior adductor muscle scar elongate, posterior adductor muscle scar subcircular; pallial sinus abrupt and rounded. **Habitat:** Inlets; offshore burrowing; infaunal at depths from 0 to 46 m (150 ft). **Remarks:** This large bivalve burrows so deep into the substratum that the dead and living shells are rarely found.

Gastrochaenidae

Gastrochaenidae is a small family of bivalves of about 15 species found in tropical seas. Shells are usually small, less than 30 mm (1⅕ in). Shells can be thin and elongate or oblong, equivalve, and broadly gaping.

Sculpture is typically of concentric growth lines. The hinge area is reduced and without dentition. Beaks are positioned anteriorly. The lower margin is convex, rarely straight; the upper margin is almost straight. The pallial sinus varies from shallow to immersed. The pallial muscles are grouped into bundles that make the pallial line discontinuous. The posterior muscle scar is larger than the greatly reduced anterior muscle scar. The foot has an almost round pedal valve that has a suctorial function to anchor the animal to the end of the bore hole or tube it inhabits. Gastrochaenids typically bore into hard, calcareous substratum. In Texas, Gastrochaenidae is represented by 1 species.

Lamychaena hians (Gmelin, 1791)

Atlantic Chimney Clam

Distribution: North Carolina to Texas; Caribbean; Bermuda; Brazil.

Size: 12 to 18 mm (½ to ¾ in).

Description: Color yellowish to whitish. Shape elongate-oval. Sculpture of coarse, commarginal, ridgelike growth lines. Umbones bulbous, turning inward, positioned at extreme narrow posterior end; area around umbones inflated with a winglike projection extending beyond the umbones. Anterior ventral margin somewhat rounded. **Habitat:** From shallow water to offshore banks; infaunal; bore into hard substrate. Found boring in rocks off East Texas coast at depths from 14 to 101 m (45 to 330 ft). **Synonym:** *Gastrochaena hians* (Gmelin, 1791).

Corbiculidae

Corbiculidae is a moderate-sized family of bivalves of about 100 species. Commonly known as marsh clams, these bivalves are found in brackish and fresh water worldwide. Shells are robust, ovate, and round to subtriangular. Sculpture may be smooth or have commarginal growth lines, and the shell is covered with a thick periostracum. Beaks are fairly large and have a distinct ligament. The internal hinge is thick and possesses up to 3 cardinal teeth in each valve. The pallial line is whole and simple; the pallial sinus is reduced or absent. The interior of the shell margin is relatively smooth. Anterior and posterior adductor muscle scars are almost equal. Corbiculids feed infaunally by pedal gape feeding. In Texas, Corbiculidae is represented by 1 genus and 2 species whose size ranges from 25 to 38 mm (1 to 1½ in); both species are discussed here.

Polymesoda caroliniana (Bosc, 1801)   **Carolina Marsh Clam**

Distribution: Virginia to Florida, Texas.
Size: 25 to 38 mm (1 to 1½ in).

Description: Color uniformly white; exterior covered with a persistent glossy brown periostracum. Shape subtriangular and broad. Sculpture of commarginal growth lines. External ligament slender; 3 cardinal teeth in each valve; 1 anterior and 1 posterior lateral tooth on either side of the cardinal teeth. Two adductor muscle scars. **Habitat:** Brackish-water marshes at depths from the surface to 2 m (7 ft). **Remarks:** This bivalve is able to withstand wide fluctuations in environmental conditions, such as salinity, anoxia, and desiccation.

Polymesoda floridana (Conrad, 1846)   **Southern Marsh Clam**

Distribution: Florida to Texas.
Size: 25 mm (1 in).

Description: Color whitish suffused with purplish-pinkish; areas on inner surface and ventral margin whitish; area beneath and around umbones purplish. Shape triangularly ovate. Sculpture smoothish with fine, commarginal growth lirae. Each valve with 3 cardinal teeth and 1 anterior and 1 posterior lateral tooth on either side of the cardinal teeth. Two adductor muscle scars. **Habitat:** Mud-sand-type bottoms. Found in hypersaline areas of the Laguna Madre at depths from low-tide line to 2 m (7 ft). **Synonym:** _P. maritima_ (d'Orbigny, 1853).

Cardiidae

Cardiidae is a relatively large family of bivalves of about 200 species commonly referred to as the heart cockles because of their heart-shaped outline in side view. Shell shape may be subcircular to triangular and occasionally squarish. It is typically inflated but may be somewhat compressed. Valves are equal to subequal. The periostracum is thin and persistent. Sculpture is usually radial with spines or projections, but some cardiids are smooth. The interior margin of the shell is typically serrate or toothed. The hinge is wide with 2 conic, curved cardinal teeth in each valve. Beaks are conspicuous. Muscle scars are oval and subequal. No pallial sinus is present, but there are openings between the mantle lobes that are bordered by thickened lobes. Cardiids are found inhabiting areas just below the surface of the sand or in sandy mud environments, intertidally to depths of 335 m (1100 ft). A strong, narrow foot allows these bivalves to vigorously move across the sandy bottom. In Texas, Cardiidae is represented by 8 genera and 13 species whose size ranges from 8 to 202 mm (⅓ to 4 in); 5 species are discussed here.

Dallocardia muricata (Linnaeus, 1758)

Yellow Prickly Cockle

Distribution: North Carolina to Florida, Texas; Caribbean; Brazil.

Size: 51 mm (2 in).

Description: Color cream to dull gray with splotches of reddish-brown; interior whitish with yellowish at periphery. Shape inflated, subcircular. Sculpture of spinose radial ribs; spines on ribs most pronounced on anterior and posterior external margins. Inner margin serrate. **Habitat:** In sand and sandy mud; along the Texas coast in hypersaline lagoons and bays. Depth range from 0 to 84 m (275 ft). **Remarks:** This cardiid replaces *Trachycardium isocardia* in the deeper parts of lagoons, bays, and offshore at depths from low-tide line to 45 m (150 ft). **Synonym:** *T. muricata* (Linnaeus, 1758).

Dinocardium robustum (Lightfoot, 1786) **Giant Atlantic Cockle**

Distribution: Virginia to Florida, Texas; Mexico.
Size: 76 to 102 mm (3 to 4 in).

Description: Color usually tannish-white on one side and brown on the other; interior usually light reddish to pale brown; anterior margin whitish, rarely found uniformly yellow. Shape inflated, broad, subquadrate. Sculpture of numerous rounded, radiating ribs crossed by concentric ribs, giving a scaly appearance. Ventral margin serrate. **Habitat:** Sand offshore (nearshore) and barrier-island beaches at depths from low-tide line to 30 m (100 ft). **Remarks:** Largest cockle on Texas coast. Reputed to make a great chowder. The form known as the "yellow cockle" is rare; when found, sometimes misidentified as a different species.

Laevicardium mortoni (Conrad, 1831) **Yellow Egg Cockle**

Distribution: Massachusetts to Florida to Texas.
Size: 18 to 25 mm (¾ to 1 in).

Description: Color a dirty white with splotches of zigzag brownish-purple; interior color variable with straw yellow base and brown to purple maculations. Shape subcircular. Sculpture of regularly spaced commarginal growth lines and fine radial threads, giving cancellate appearance; gentle sulcus from umbo to posterior ventral angle. **Habitat:** Open, shallow, hypersaline lagoons close to inlets at depths from shoreline to 84 m (275 ft). **Remarks:** _L. mortoni_ is common in beach drift, particularly on South Texas beaches. This cockle is able to swim by sculling its extended foot.

Laevicardium serratum (Linnaeus, 1758)　　　　　　　　**Egg Cockle**

Distribution: North Carolina, Florida, Texas; Caribbean; Bermuda.

Size: 25 to 51 mm (1 to 2 in).

Description: Color of exterior variable, white to rose to orange; interior coloration typically same as external color. Shape elongate-oval to quadrate. Sculpture smooth, with fine radial threads and concentric threads, giving fine, latticelike appearance; exterior sculpture shows through to interior; smooth, dorsal, central margin on both sides of umbo. Ventral margin crenulate. **Habitat:** Muddy bottoms at depths from low-tide line to 75 m (246 ft). **Remarks:** Formerly known as _L. laevigatum_ (Linnaeus, 1758), which is from the Indo-Pacific. Changes in shell shape at different sizes are a characteristic feature of this bivalve.

Trachycardium isocardia (Linnaeus, 1758)　　　　　**Prickly Cockle**

Distribution: Texas; Caribbean; Bermuda.

Size: 76 mm (3 in).

Description: Color tannish with large shaded areas of yellowish-brown; interior glossy rose with white near the anterior and posterior margins, yellow on serrations of inner ventral margin. Shape inflated, ovate. Sculpture of 31 to 37 highly scaled ribs with thickened spines on ribs of posterior sulcus. Inner margin serrate. **Habitat:** Grass beds and sandy bottoms. Depth range from 0 to 37 m (121 ft). **Remarks:** This species is located commonly in the shallow seagrass beds of South Padre Island (R. Davenport, pers. comm., 2005; J. Nill, pers. comm., 2007). Turgeon et al. (1998) state that this species is not found in the continental United States.

Veneridae

Veneridae is a large family of bivalves of about 500 species found world-wide in marine environments and sometimes brackish water. The shell is usually thick and varies in shape from ovate or elongate to subcircular, subtriangular, or subquadrate. A variety of sculpture is represented in the venerids; the shells can be smooth with strong or weak concentric ridges, radial ribs, or both concentric and radial ribs, giving a cancellate appearance, sometimes with spines and/or scales. Beaks are large and located anterior to the center of the shell. The hinge plate is broad and usually equipped with 3 internal cardinal teeth in each valve; lateral teeth are typically reduced or absent. The pallial sinus varies in size in different species, and the pallial line is complete. The posterior muscle is either larger than or equal in size to the anterior muscle. The interior ventral margin may be smooth or crenulate. The foot is strong enough to allow for efficient burrowing into sand or mud. The venerids possess well-developed and separate siphons. Some venerids are good to eat and are sold commercially. In Texas, Veneridae is represented by 20 genera and 29 species whose size ranges from 5 to 152 mm (⅕ to 6 in); 16 species are discussed here.

Agriopoma texasiana (Dall, 1892) — Texas Venus

Distribution: Florida to Texas; Mexico.

Size: 38 to 76 mm (1½ to 3 in).

Description: Creamy white to gray; interior chalk white. Shape ovate. Sculpture absent except for faint commarginal growth lines. Umbones rolled inward; lunule weak; posterior tooth S-shaped in each valve. Pallial line distinct with a triangular sinus that touches the posterior adductor muscle scar. Inner ventral margin smooth. **Habitat:** Burrowed in clay environments such as lagoons and offshore at depths from 7 to 24 m (23 to 80 ft). **Remarks:** Formerly placed in the genus *Callocardia*. Dead shells are commonly collected in beach drift all along the Texas coast. **Synonym:** *C. texasiana* (Dall, 1892).

Anomalocardia auberiana (d'Orbigny, 1853)   **Pointed Venus**

Distribution: Florida, Texas; Quintana Roo.
Size: 10 mm (¾ in). Specimen in photograph, 6 mm (¼ in).

Description: Color variable, from white to tannish-brown suffused with brown. Shape triangular and inflated. Sculpture with broad commarginal ribs and smaller threads in inner spaces; sulcus from umbo to posterior ventral angle.
Habitat: Hypersaline bays and lagoons of the Texas coast; Laguna Madre at depths from 0 to 4 m (13 ft). **Remarks:** _A. auberiana_ is a significantly smaller species than _A. cuneimeris;_ thus, some authors consider it a synonym of _A. cuneimeris._ Specimen in photograph collected in the Upper Laguna Madre seagrass beds; from Tunnell collection.

Anomalocardia cuneimeris (Conrad, 1846)   **Pointed Venus**

Distribution: Florida to Texas.
Size: 12 to 18 mm (½ to ¾ in).

Description: Color variable, from white to tannish-brown suffused with brown or bluish-gray splotches; inner surface suffused with purplish-brown and margins white. Shape triangularly ovate and inflated. Sculpture with concentric ridges crossed by faint radial ribs; sulcus extends from umbo to posterior ventral angle.
Habitat: Most common in hypersaline bays and lagoons of the South Texas coast. Depth range is from 0 to 0.3 m (1 ft). **Remarks:** Irregular concentric sculpture varies in number, and overall outline may change from triangular to wedge shaped.

Chione elevata (Say, 1822)   **Florida Cross-barred Venus**

Distribution: North Carolina to Florida, Texas; Caribbean; Brazil.

Size: 25 to 45 mm (1 to 1¾ in).

Description: Color variable from white to grayish-white with brownish spots; inner margins whitish with central area usually suffused with purplish-brown. Shape ovately triangular. Sculpture of wavy, platelike concentric ribs crossed by radial ribs; periphery of escutcheon angled. **Habitat:** Sandy bottoms from below low-tide line to a depth of about 101 m (330 ft). **Remarks:** Formerly misidentified as *C. cancellata* (Linnaeus, 1767). A recent study using morphological, morphometric, and phylogenetic analysis of shell characters determined the two to be separate and unique, with *C. elevata* found in northern areas (including Texas) and *C. cancellata* found tropically in the Caribbean.

Choristodon robustum (Sowerby I, 1834)   **Boring Choristodon**

Distribution: North Carolina to Florida, Texas; Brazil.

Size: 25 mm (1 in).

Description: Color uniformly yellowish-white. Shape elongate-triangular. Sculpture of strong, somewhat narrow, rounded, irregularly spaced radial ribs crossed by concentric ribs, giving a beaded look where they intersect. Umbones placed anteriorly. Anterior end wide and round, posterior end narrower. Pallial sinus larger than posterior adductor muscle scar. **Habitat:** Borers in hard substrate. In Texas found alive at depths from 11 to 22 m (36 to 72 ft). Depth range from 1 to 38 m (125 ft). **Remarks:** Typically a shallow-water species found "in" rocks or other hard substrates. **Synonym:** *Rupellaria typica* (Jonas, 1844).

Cyclinella tenuis (Récluz, 1852)   **Atlantic Cyclinella**

Distribution: Virginia to Texas to Brazil.
Size: 18 to 25 mm (¾ to 1 in).

Description: Color uniformly yellowish-white. Shape sub-circular, somewhat inflated. Sculpture of irregular com-marginal growth lines. Umbones positioned slightly anteriorly, recurved, small-ish; ligament of medium size and immersed; lunule long; escutcheon obscure. Pallial sinus long, circular-elongate, and directed dorsally. **Habitat:** Shallow, sandy mud bottoms in bays and on beaches. In Texas commonly collected at depths from shoreline to 1 m (3 ft). Depth range from 0 to 66 m (217 ft).

Dosinia discus (Reeve, 1850)   **Disk Dosinia**

Distribution: Virginia to Florida, Texas; Bahamas.
Size: 51 to 76 mm (2 to 3 in).

Description: Color a glossy yellowish-white; inner surface white. Shape subcircular, compressed. Sculpture of numerous, crowded concentric ribs. Lunule impressed, well defined; escutcheon obscure. Pallial apparatus well defined. Inner margin smooth. **Habitat:** On protected beaches in sand or sandy mud in intertidal zone. Depth range from 0 to 79 m (260 ft). **Remarks:** Common beach species found washed ashore. Because of this species' strong ligamental structure, valves are typically found held together.

Dosinia elegans (Conrad, 1843)   **Elegant Dosinia**

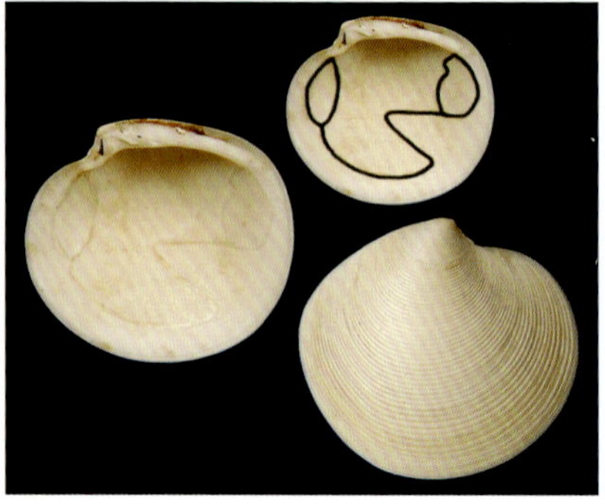

Distribution: North Carolina to Texas; Caribbean.
Size: 51 to 76 mm (2 to 3 in).

Description: Color glossy yellowish-white, rarely with mottlings of reddish-brown; inner surface whitish-yellow and suffused with pinkish-purple usually beneath the umbones. Shape subcircular. Sculpture of crowded concentric ridges. Pallial apparatus well defined. Umbones recurved, arched forward. **Habitat:** In sand or sandy mud at depths from 0 to 79 m (260 ft). **Remarks:** *D. discus* and *D. elegans* are easily confused but can be distinguished by *D. discus* being thinner shelled and more compressed and having more concentric ribs. *D. elegans* replaces *D. discus* in deeper water.

Gemma gemma (Totten, 1834)   **Amethyst Gem Clam**

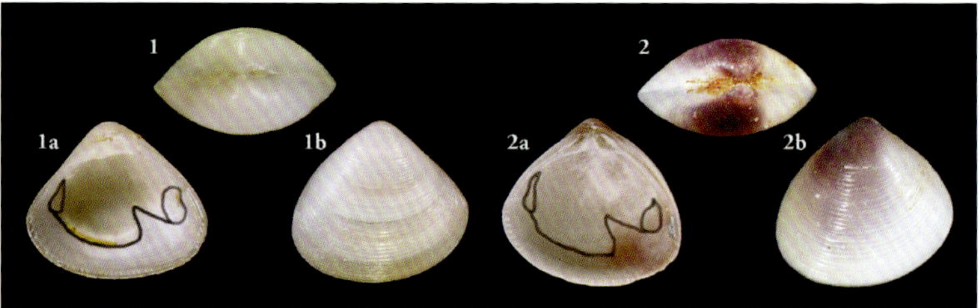

Distribution: Nova Scotia to Florida, Texas; Bahamas.
Size: 3 mm (⅛ in).

Description: Color variable from whitish to whitish with purplish coloration on umbones or suffused into the shell; inner surface polished whitish to whitish-bluish-purple. Shape triangular. Sculpture smoothish with fine, closely spaced commarginal ribs. Umbones distinct; 2 cardinal teeth in left valve with a socket in between, 3 cardinal teeth in right valve. Ventral margin crenulate in fresh material. **Habitat:** Infaunal. Sandy mud bottoms. Depth range from 0 to 66 m (217 ft). **Remarks:** This species has been collected alive in mudflats around Port Aransas. Specimen 1: whitish form; specimen 2: with purplish umbones. Specimens with purple coloration originally named *G. purpurea* (Lea, 1842). **Synonym:** *G. purpurea* (Lea, 1842).

Lirophora clenchi (Pulley, 1952)

Clench's Thick-ringed Venus

Distribution: Texas to Gulf of Campeche, Mexico.
Size: 25 to 34 mm (1 to 1⅓ in).

Description: Color of base white with strong brown maculations; inner surface white, stained brown toward posterior inner margin. Shape irregularly triangular. Sculpture of strong, well-developed, broad, rounded, irregularly shaped, wavy cords with numerous fine concentric threads irregularly spaced above the cords. Umbones curved forward and inward. **Habitat:** Sand-mud-type bottoms and shale uplifts. In Texas found alive at depths from 12 to 50 m (42 to 168 ft). Depth range from 0 to 91 m (300 ft). **Remarks:** Commonly collected dead along the South Texas beaches. **Synonym:** *Chione clenchi* (Pulley, 1952).

Macrocallista nimbosa (Lightfoot, 1786)

Sunray Venus

Distribution: North Carolina to Florida to Texas.
Size: 102 to 150 mm (4 to 6 in).

Description: Color reddish to salmon, glossy with darker interrupted bands of rays from umbones to ventral margin; inner surface dull white with a reddish blush suffused in central area of shell of fresh specimens. Shape elongate-oval. Sculpture smoothish with only irregular fine commarginal growth lines. Lunule elongate and impressed; escutcheon bordered by a color change; hinge elongate and slender. **Habitat:** Sandy bottom. **Remarks:** Easily recognizable because no other venerid reaches its size or coloration.

Mercenaria campechiensis (Gmelin, 1791)   **Southern Quahog**

Distribution: New Jersey to Florida, Texas; Cuba.
Size: 76 to 152 mm (3 to 6 in).

Description: Color dull grayish-white, stained brown on margins; inner surface white. Shape subcircular. Sculpture of numerous thick, concentric ribs. Well-developed lunule, almost as high as wide; escutcheon somewhat well developed with a ridge where concentric growth lines change direction. Small angular pallial sinus. **Habitat:** Calcareous environments and sandy bottoms offshore at depths from 0 to 60 m (200 ft). **Remarks:** Similar to *M. texana* (Dall, 1902), but sculpture in *M. texana* is weaker and fades out toward the center of the outer surface of the shell. The lunule is larger and shell is heavier and larger in *M. campechiensis.*

Mercenaria texana (Dall, 1902)   **Texas Quahog**

Distribution: Texas, northern Gulf of Mexico.
Size: 76 to 152 mm (3 to 6 in).

Description: Color dull grayish-white; inner surface white; periostracum brown. Shape subcircular. Sculpture of numerous weak, concentric ribs. Well-developed lunule almost as high as wide; escutcheon somewhat well developed with a ridge where the concentric growth lines change direction; 3 cardinal teeth in each valve. Small angular pallial sinus. **Habitat:** Bays and inlets of Texas. Depth range from 0 to 1 m (3 ft). **Remarks:** Similar to *M. campechiensis,* but *M. campechiensis* is sculptured throughout the surface of the valve. Overall weight and height are less and lunule is smaller in *M. texana.* **Synonym:** *M. campechiensis texana* (Dall, 1902).

Petricolaria pholadiformis (Lamarck, 1818)   **False Angelwing**

Distribution: Gulf of St. Lawrence to Texas; Uruguay. **Size:** 51 mm (2 in).

Description: Color uniformly white. Shape elongate-oval, cylindrical. Sculpture with large radial ribs at anterior end and rest of shell with numerous fine, radial ribs; concentric ribs present, becoming scaly and cancellate where they cross the larger, anterior radial ribs. Umbones positioned anteriorly; pallial sinus slender and deep. Apophysis absent. **Habitat:** Bays, inlets, near shore in clay-type environments. Depth range from 0 to 49 m (160 ft). **Remarks:** Common species found along the Texas coast. Resembles _Pholas;_ however, analysis of its dentition quickly corrects the error. Formerly in family Petricolidae (now a subfamily in Veneridae). **Synonym:** _Petricola pholadiformis_ (Lamarck, 1818.)

Puberella intapurpurea (Conrad, 1849)   **Lady-in-Waiting Venus**

Distribution: North Carolina to Texas; Caribbean; Brazil. **Size:** 25 to 38 mm (1 to 1½ in).

Description: Color dull gray with splotches of reddish-brown; inner surface suffused with purplish-red. Shape ovately triangular, inflated. Sculpture of distinct, regularly spaced, concentric, plate-like ridges with radial riblets in inner spaces; concentric ribs give serrate appearance. Lunule heart shaped with raised, leaflike lamellations; escutcheon with thin transverse threads. Ventral margin crenulate posterior to dorsal margin. **Habitat:** Sandy, shelly bottoms. On Texas beaches and in bays. Found alive at depths from 11 to 22 m (36 to 72 ft). Depth range from 0 to 55 m (180 ft). **Synonym:** _Chione intapurpurea_ (Conrad, 1849).

Timoclea grus (Holmes, 1858)   **Gray Pygmy Venus**

Distribution: North Carolina to Florida to Texas.

Size: 6 to 11 mm (¼ to 3/8 in).

Description: Color grayish-tan, sometimes with purplish stains on posterior slope; inner surface white, sometimes stained brown. Shape subquadrate. Sculpture of regularly spaced, somewhat raised concentric ribs crossed by radial ribs, giving cancellate appearance. Lunule long and brown; escutcheon slender; 3 compressed cardinal teeth in each valve. **Habitat:** On shelly or rocky bottoms. In Texas found alive at depths of about 12 to 54 m (42 to 180 ft). Depth range from 0 to 101 m (330 ft). **Remarks:** Widespread species found all along the Louisiana-Texas offshore coast but rarely in bays. **Synonym:** *Chione grus* (Holmes, 1858).

Tellinidae

Tellinidae is a large family of bivalves of approximately 350 species found throughout the world; however, the majority are found in tropical seas in shallow-water infaunal habitats. Many are important food sources, and because of their diverse coloration the majority of tellins are attractive to shell collectors. Shells are primarily flattened and ovate, but some are subcircular, triangular, or oblong. The anterior end is typically rounded, and the posterior end is elongate. The left valve is usually deeper and larger than the right. The right valve typically has a ridge extending from the beak to the ventral angle, and the left valve has a corresponding groove. Tellins typically lie on their left valve, thus allowing the right valve to become more inflated and convex. The majority of tellin sculpture is smooth; however, some tellins display concentric ridges and radial riblets. Muscle scars are present, and the pallial apparatus is distinct. In Texas, Tellinidae is represented by 9 genera and 25 species whose size ranges from 8 to 114 mm (⅓ to 4½ in); 16 species are discussed here.

Angulus tampaensis (Conrad, 1866)

Distribution: Florida to Texas; Caribbean.

Size: 12 to 25 mm (½ to 1 in).

Description: Color whitish, sometimes tinged with pink or orange; interior white. Shape broadly ovate to subtriangular. Sculpture of irregular commarginal growth lines and finer concentric threads between. Anterior margin broadly rounded; posterior margin somewhat truncate. Each valve with 2 cardinal teeth. **Habitat:** Usually in sand at depths from shoreline to 1 m (3 ft). **Remarks:** In Texas _A. tampaensis_ can be located in water shallower than 150 mm (6 in) in the hypersaline lagoons and tidal flats of South Texas bays and lagoons. **Synonym:** _Tellina tampaensis_ (Conrad, 1866.)

Angulus texanus (Dall, 1900) Texas Tellin

Distribution: North Carolina to Texas; Mexico to Bahamas.

Size: 12 mm (½ in).

Description: Color usually uniformly white, sometimes stained with yellow with some iridescence on the outer surface and highly polished internally. Shape elongate-oval. Sculpture smoothish of fine concentric threads that sometimes become obsolete. Anterior margin long and gently rounded; posterior margin truncate. Each valve with 2 cardinal teeth. **Habitat:** In mud bottoms. A bay species that occupies both low-salinity eastern bays and high-salinity southern bays of Texas. Depth range from 0 to 46 m (150 ft). **Remarks:** One of the only tellins in Texas able to tolerate wide ranges of salinity. **Synonym:** _Tellina texana_ (Dall, 1900.)

Angulus versicolor (DeKay, 1843)

Many-colored Dwarf Tellin

Distribution: Rhode Island to Florida to Texas; Caribbean.
Size: 12 mm (½ in).

Description: Color variable, translucent whitish to red to pink with iridescence on outer surface and glassy internally. Shape somewhat inflated, sub-elliptical. Sculpture fragile with fine, irregularly spaced com-marginal growth lines. Umbones located posteriorly, small and acute. Anterior dorsal margin long and gently sloping; posterior dorsal margin inclined and truncate with a sulcus from umbo to ventral margin. **Habitat:** In sand and sandy mud bottoms of Texas bays and beaches. In Texas found alive at depths from shoreline to about 20 m (66 ft). Depth range from 0 to 46 m (150 ft). **Synonym:** *Tellina versicolor* (DeKay, 1843).

Eurytellina alternata alternata (Say, 1822)

Alternate Tellin

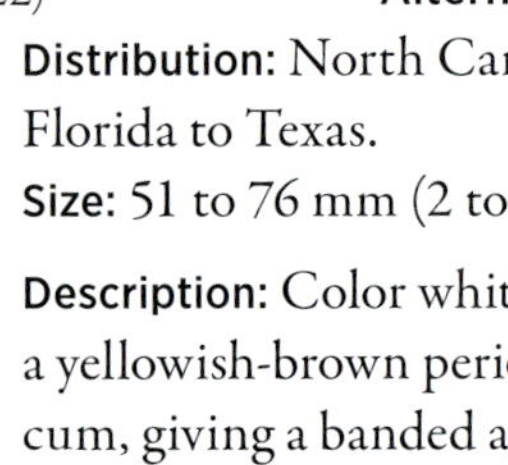

Distribution: North Carolina, Florida to Texas.
Size: 51 to 76 mm (2 to 3 in).

Description: Color whitish with a yellowish-brown periostracum, giving a banded appearance; internally yellowish with white margins. Shape elongate-oval. Sculpture of incised concentric grooves; slight ridge on anterior dorsal margin, and a narrow sulcus on the posterior end that extends from umbo to ventral angle. Internal ridge in both valves extends from umbo to anterior adductor muscle scar. **Habitat:** Sandy substrate at depths from 0 to 128 m (420 ft). **Remarks:** Similar to *E. a. tayloriana,* but *E. a. tayloriana* has a bright pink shell and a narrower internal cavity. **Synonym:** *Tellina alternata* (Say, 1822).

Eurytellina alternata tayloriana (Sowerby II, 1867) **Taylor's Alternate Tellin**

Distribution: Texas to Tampico, Mexico.

Size: 51 to 76 mm (2 to 3 in).

Description: Color different shades of pink; internally bright pink with suffusions of white. Shape elongate-oval. Sculpture of incised concentric grooves with a slight ridge on the anterior dorsal margin and a narrow sulcus on the posterior end that extends from umbo to ventral angle. Internal ridge in both valves extends from umbo to anterior adductor muscle scar. **Habitat:** Sandy substrate at depths from 0 to 128 m (420 ft). **Remarks:** Similar to *E. a. alternata* but differs ecologically in that *E. a. alternata* is typically found at greater depths and the right valve is narrower. **Synonym:** *Tellina alternata tayloriana* (Sowerby II, 1867).

Eurytellina lineata (Turton, 1819) **Rose-petal Tellin**

Distribution: Florida to Texas to Brazil.

Size: 38 mm (1½ in).

Description: Color of outer surface white, tinged with yellow or rose on umbones; inner surface rose tinged with white. Shape broadly ovate. Sculpture of concentric growth lines and radial wrinkles; posterior with sulcus that extends from umbo to ventral angle. Adductor muscle scars well impressed. Pallial sinus does not rise above adductor muscles. Each valve with 2 cardinal teeth and 2 lateral teeth; posterior lateral in left valve obsolete. **Habitat:** Offshore reefs and banks. Shells collected from shoreline to 45 m (150 ft). **Remarks:** Loose valves have been collected in beach drift and spoil from South Texas bays. Live material rare. **Synonym:** *Tellina lineata* (Turton, 1819).

Macoma brevifrons (Say, 1834)   Short Macoma

Distribution: South Carolina, Texas; Brazil.
Size: 38 mm (1½ in).

Description: Color uniformly whitish tinged with peach; outer exterior margin with a dark brown narrow band. Shape elongate-oval. Sculpture of commarginal growth lines and fine radial lirae. Each valve with 2 central teeth; lateral teeth absent. Anterior adductor muscle scar slender and irregular. Anterior end shorter and rounder.
Habitat: Typically found near shore and possibly in bays. Depth range from 0 to 11 m (36 ft). **Remarks:** *M. brevifrons* and *Angulus tampaensis* are similar in size and color but differ in the hinge area.

Macoma constricta (Bruguière, 1792)   Constricted Macoma

Distribution: Florida to Texas; Caribbean; Brazil.
Size: 25 to 38 mm (1 to 1½ in).

Description: Color uniformly white with brown periostracum. Shape obliquely oblong. Sculpture smoothish with irregular commarginal growth lines. Rounded anteriorly; a noticeable ridge extends from umbo to narrower posterior ventral angle. Each valve with 2 cardinal teeth, but posterior cardinal tooth in left valve typically small to obsolete; lateral teeth absent.
Habitat: Typically a bay bivalve. Usually resides in sandy mudflats in shallow water on Texas bay shores. Found alive at depths from shoreline to 2 m (7 ft). **Remarks:** One of the more eurytypic (can tolerate wide ranges) species of macomas on the Texas coast.

Macoma mitchelli (Dall, 1895)   Matagorda Macoma

Distribution: South Carolina to Texas.

Size: 15 mm (⅗ in).

Description: Color uniformly iridescent whitish. Shape elongate-oval, squarish anteriorly and truncate posteriorly. Sculpture with fine commarginal growth lines. Each valve with 2 cardinal teeth; posterior cardinal teeth in left valve obsolete; lateral teeth absent. **Habitat:** River-influenced, brackish to almost fresh water. Shells found at depths from shoreline to 36 m (120 ft). In Texas alive at depths from shoreline to 2 m (7 ft). Depth range from 0 to 36 m (120 ft). **Remarks:** Common in Matagorda and Galveston Bays. Lives in association with *Rangia* species.

Macoma pulleyi (Boyer, 1969)   Delta Macoma

Distribution: West of Mississippi Delta, Louisiana, Texas.

Size: 25 to 51 mm (1 to 2 in).

Description: Color uniformly white. Shape elongate-rectangular. Sculpture with commarginal growth lines; umbones point posteriorly. Anterior end longer and squarish; posterior end diagonally truncate. Sulcus extends from umbo to posterior ventral angle; ventral margin almost straight. Hinge narrow; both valves possess a bifurcate cardinal tooth. **Habitat:** Offshore in dark blue clay at depths from 1 to 128 m (3 to 420 ft). **Remarks:** *M. pulleyi* is often confused with *M. tageliformis;* however, *M. pulleyi* has a steeper, more abrupt posterior angle. The sulcus bears a distinct ridge that delineates it from the rest of the shell.

Macoma tageliformis (Dall, 1900)
Tagelus-like Macoma

Distribution: Louisiana to Texas; Greater Antilles; Brazil.
Size: 25 to 51 mm (1 to 2 in).

Description: Color uniformly white. Shape elongate-oval. Sculpture with commarginal growth lines. Umbones point posteriorly. Anterior end longer and rounded; posterior end diagonally truncate. Sulcus extends from umbo to posterior ventral angle. Each valve with 2 cardinal teeth, but the posterior cardinal tooth in left valve small to obsolete. **Habitat:** An infaunal species of the surf zone. Shells found at depths from 1 to 128 m (3 to 420 ft). **Remarks:** *M. tageliformis* is often confused with *M. pulleyi;* however, *M. tageliformis* is rounder anteriorly, *M. pulleyi* is squarish, and the pallial sinus is wider and shallower in *M. tageliformis.*

Macoma tenta (Say, 1834)
Elongate Macoma

Distribution: Cape Cod, Massachuetts to Florida, Texas; Brazil; Bermuda.
Size: 12 to 18 mm (½ to ¾ in).

Description: Color translucent white; periostracum grayish. Shape elongate-oval. Sculpture smooth with fine, barely visible commarginal growth lines that show through interiorly. Anterior end long and rounded; posterior end with a gentle ridge from umbo to posterior ventral angle. Umbones point posteriorly; right valve with 2 cardinal teeth; left valve with 1 cardinal tooth. **Habitat:** Fairly common throughout the coastal bays of Texas at depths from 0 to 12 m (40 ft). **Remarks:** *M. tenta* is distinguished from *M. extenuata* by its iridescence and pallial sinus, which is rounded above.

Merisca aequistriata (Say, 1824)   **Striate Tellin**

Size: 18 to 25 mm (¾ to 1 in).

Description: Color uniformly white. Shape somewhat oval. Sculpture of regular concentric ribs. Umbones point posteriorly. Rounded anteriorly; truncate posteriorly with a slight sulcus that extends from umbo to posterior ventral angle. Each valve with 2 cardinal teeth; well-developed lateral teeth in right valve; lateral teeth obsolete in left valve. Pallial sinus irregular, joins pallial line below anterior adductor muscle scar. **Habitat:** Sandy bottom offshore. Widespread species found all along the Texas shore at depths from 0 to 91 m (300 ft). **Synonym:** _Tellina aequistriata_ (Say, 1824).

Scissula iris (Say, 1822)   **Iris Tellin**

Distribution: North Carolina to Florida to Texas.

Size: 12 mm (½ in).

Description: Color translucent white or rose. Shape elongate-oval. Sculpture fragile with concentric threads intersected by scissulations (having sculpture running obliquely across the shell) at approximately 25°. Anterior adductor muscle scar quadrate; posterior adductor muscle scar rounded. Umbones small, placed posterior to middle of shell. Each valve with 2 cardinal teeth; left anterior and right posterior cardinal teeth bifurcate; right valve with lateral tooth proximal to the cardinal complex. **Habitat:** Sandy beaches and inlet mudflats at depths from 0 to 37 m (121 ft). **Remarks:** In its habitat _S. iris_ is usually the least noticed but most common bivalve. **Synonym:** _Tellina iris_ (Say, 1822).

Strigilla mirabilis (Philippi, 1841) — White Strigilla

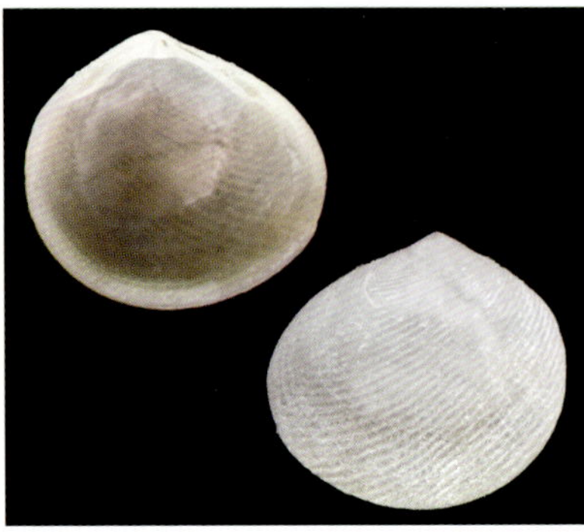

Distribution: Bermuda; North Carolina to Texas; Caribbean; Brazil.

Size: 8 to 12 mm (⅓ to ½ in).

Description: Color translucent white. Shape inflated, subcircular. Sculpture of concentric growth lines crossed by oblique ribs that meet the ventral margin at about a 45° angle, giving diamondlike, cancellate appearance; posterior slope with about 4 zigzag lines. Umbones round, almost centrally located. Each valve with 2 cardinal teeth, but posterior cardinal in left valve small to obsolete. **Habitat:** In sandy mud, shale outcrops, and coral debris at depths from 0 to 60 m (200 ft). **Remarks:** Although uncommon in Texas, this species is easily recognized by its typical oblique sculpture.

Tellidora cristata (Récluz, 1842) — White-crested Tellin

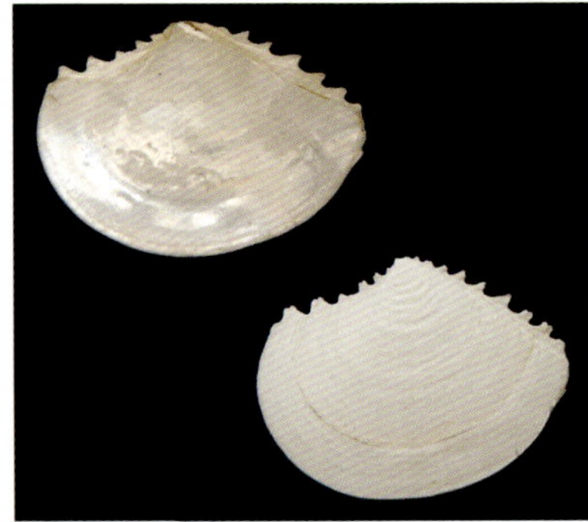

Distribution: North Carolina to Florida, Texas.

Size: 25 to 38 mm (1 to 1½ in).

Description: Color chalk white; interior iridescent. Shape somewhat triangular. Sculpture of irregular concentric ribs; dorsal margin with distinct projecting spines; spines form deep lunule and escutcheon. Each valve with 2 cardinal teeth, with strong anterior lateral tooth in right valve. Double adductor muscle scars with anterior adductor muscle scar longer than posterior adductor muscle scar. **Habitat:** Inlets, channels, and bay margins at depths from 0 to 46 m (150 ft). **Remarks:** The spines located on the dorsal margin make this an easily identifiable species.

Donacidae

Donacidae is a relatively small family of bivalves of approximately 50 species that inhabit all warm marine waters. Shells are typically heavy and triangular to wedge shaped, and shells of some species display a variety of coloration. Sculpture is smooth but may have radial or slanting striae. Beaks are relatively large and positioned posteriorly on the shell. The posterior end is typically broader than the narrower anterior end. Often an angled ridge will be positioned from the beaks to the lower margin, producing a posterior slope. The interior portion of the hinge on each valve has 2 cardinal teeth and a lateral tooth on each side. The shell margin is crenulate and furrowed. Unlike the other tellinoideans, the donacids are clearly defined by their shell. These bivalves are characteristic of sandy beaches, where they bury themselves just beneath the surface. Their distribution in life is vertical. When uncovered by wave action, *Donax* can rebury themselves efficiently and quickly. In Texas, Donacidae is represented by 1 genus and 2 species whose size ranges from 12 to 25 mm (½ to 1 in); both species are discussed here.

Donax texasianus (Philippi, 1847) — Texas Coquina

Distribution: Louisiana, Texas to Veracruz, Mexico.

Size: 5 to 8 mm (⅕ to ⅓ in).

Description: Color variable, whitish with light blue, pink, or yellowish suffusions, rarely rayed. Shape triangular, somewhat inflated. Sculpture smoothish with commarginal growth lines and radial threads with 2 cardinal teeth in each valve. Pallial sinus rounded. Ventral posterior margin crenulate; ventral anterior margin smooth. **Habitat:** Sandy beaches; intermediate shallow surf zone. Intertidal to shallow subtidal; minimal migration tidally.

Donax variabilis (Say, 1822)   **Variable Coquina**

Distribution: New York to Florida, Texas.

Size: 15 to 20 mm (⅗ to ⅘ in).

Description: Color variable (as the name implies), usually rayed with white, pink, purple, bluish, or mauve. Shape elongate-triangular. Sculpture smoothish with commarginal growth lines and irregular radial threads. Cardinal and lateral teeth present. Ventral posterior margin crenulate; ventral anterior margin smooth. **Habitat:** Surf zone of sandy beaches; typically migrate with tide. Depth range from 0 to 11 m (36 ft). **Remarks:** Although they spend most of their time within the sand, they periodically emerge to catch a wave to migrate along the beach. See video by Jace Tunnell: https://youtu.be/6qtAyMeIYsQ.

Psammobiidae

Psammobiidae is a moderate-sized family of bivalves of about 100 species. They are found throughout the world in tropical and temperate waters. These bivalves are similar to the tellins but differ in not having a flexure or lateral teeth. There have been questions on the validity of separating these bivalves from the tellins. Shell shape is elongate to oval. Sculpture is of smooth to weak commarginal lines. The hinge plate has 1 to 3 cardinal teeth. Lateral teeth are absent. The pallial sinus is distinct, and the pallial line is well defined. Most psammobiids are infaunal deposit feeders. Habitats vary; some psammobiids are found in "clean" sand, and others are found in mud. In Texas, Psammobiidae is represented by 2 genera and 2 species whose size ranges from 38 to 51 mm (1½ to 2 in); 1 species is discussed here.

Sanguinolaria sanguinolenta (Gmelin, 1791) **Atlantic Sanguin**

Distribution: Florida, Texas; Caribbean; Brazil.

Size: 38 to 51 mm (1½ to 2 in).

Description: Color uniformly white, except reddish-orange area beneath umbones when fresh. Shape elongate-oval. Sculpture of microscopic growth lirae. Left valve slightly more compressed than right valve. Hinge line with 2 small cardinal teeth in each valve. Adductor muscle scars placed toward dorsal margin. Pallial sinus distinct and deep. **Habitat:** Infaunal. Sandy bottoms; depth uncertain. Depth range from 0 to 1 m (3 ft). **Remarks:** Any coloration this burrowing clam shows fades quickly once being washed ashore. It is thought to be adventitious and not a resident of Texas waters.

Semelidae

Semelidae is a relatively small family of bivalves made up of approximately 60 species. These bivalves are found in tropical and subtropical waters throughout the world. The semeles are similar to the tellins. The anterior end is rounded, and the posterior end is truncate, beaklike, or subcircular. The beaks are generally located centrally but sometimes are found posteriorly. Sculpture is predominantly smooth but may be concentric, radial, or both. The hinge area usually has 2 obscure cardinal teeth, and lateral teeth are usually present in both valves. The muscle scars are subequal. The pallial sinus is round and considerable, and the pallial line is distinct. The large foot is flattened and nonbyssate. In Texas, Semelidae is represented by 5 genera and 10 species whose size ranges from 5 to 38 mm (⅕ to 1½ in); 6 species are discussed here.

Abra aequalis (Say, 1822)   **Common Atlantic Abra**

Distribution: North Carolina to Texas; Caribbean.
Size: 6 mm (¼ in).

Description: Color white with some iridescence; internal surface glassy white. Shape subcircular, inflated. Sculpture smooth with obscure, fine commarginal growth lirae. Anterior margin smoothly rounded; posterior margin subquadrate. Right valve with 2 cardinal teeth and 1 distinct lateral tooth; left valve with 1 cardinal tooth and no lateral teeth.
Habitat: Texas bays, inlets, beaches, and shallow water to a depth of 73 m (240 ft). **Remarks:** There are 2 forms, a thick-shelled one and a thin, shinier, shelled one. Commonly dredged from shallow offshore shelf.

Cumingia tellinoides (Conrad, 1831)   **Tellin Semele**

Distribution: Nova Scotia to St. Augustine, Florida, Texas.
Size: 10 to 20 mm (⅖ to ⅘ in).

Description: Color uniformly white. Shape broadly ovate. Sculpture of fine, low, irregularly spaced concentric ridges and faint to obscure radial wrinkles; gentle ridge extends from dorsal posterior margin to posterior ventral angle. Right valve has 2 cardinal teeth in front of ligamental pit and 1 lateral tooth on either side of resilium; left valve with 2 cardinal teeth fused posteriorly, lateral teeth obscure. **Habitat:** In mud or sand at depths from shoreline to 68 m (223 ft). **Remarks:** Density of concentric ribbing is variable but easily reaches numbers greater than those of _Cumingia coarctata_ (Sowerby I, 1833) (see Tunnell et al. 2010: 376).

Ervilia concentrica (Holmes, 1860)   **Concentric Ervilia**

Distribution: North Carolina to Florida, Texas; Brazil; Bermuda.

Size: 5 to 8 mm (⅕ to ⅓ in).

Description: Color whitish. Shape elliptical to triangular. Sculpture of numerous fine, irregular concentric growth lines. Anterior end abrupt and rounded; posterior end rostrate. Umbones located anteriorly; beaks point posteriorly; right valve has a large cardinal tooth; left valve has a pit or resilium to accommodate the tooth. **Habitat:** Common on shell banks of Texas. Shells found at depths from 0 to 100 m (330 ft). **Remarks:** *E. concentrica* shape varies. Ribbing is sometimes more pronounced, and its umbo is sometimes more centrally located than in its usual anterior position.

Semele bellastriata (Conrad, 1837)   **Cancellate Semele**

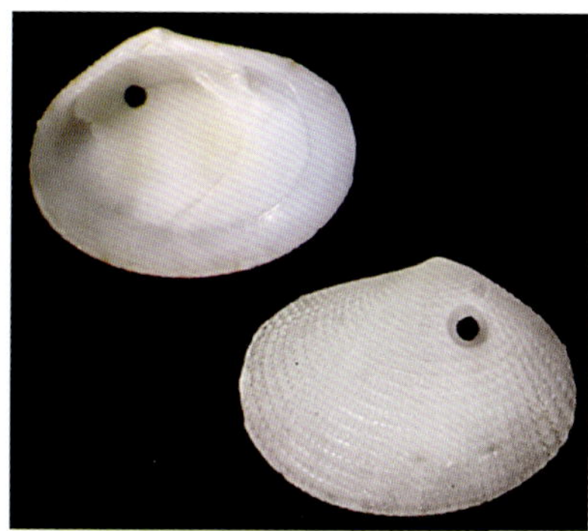

Distribution: North Carolina to Florida, Texas; Caribbean; Brazil; Bermuda.

Size: 12 to 18 mm (½ to ¾ in).

Description: Color from white to almost purple, tinges of purple typically found radially or peripherally; inner surface white with occasional suffusions of yellow. Shape oval to subrectangular. Sculpture of distinct concentric ridges crossed by radial ribs, giving cancellate appearance. **Habitat:** Sandy mud bottoms along the Texas coast and Gulf of Mexico. Shells found in depths from low-tide line to 90 m (300 ft). Live material found at depths from low-tide line to 47 m (156 ft). **Remarks:** Surface sculpture shows variation from predominantly concentric ridges and a few radial ribs to strongly cancellate.

Semele proficua (Pulteney, 1799)   **White Atlantic Semele**

Distribution: North Carolina to Florida, Texas to Brazil; Bermuda.

Size: 13 to 38 mm (½ to 1½ in).

Description: Color grayish-white; interior white sometimes suffused with yellow; slender lunule light reddish-brown. Shape subcircular. Sculpture of fine concentric growth lines and microscopic radial wrinkles strongest at ventral margin. Beaks somewhat pointing forward. Anterior and ventral margins broadly rounded; posterior weakly convex. **Habitat:** Sandy areas of bays, inlets, and shallow offshore waters at depths from 0 to 75 m (246 ft). **Remarks:** An important variation of the shell is its coloration. Sometimes it is only tinted slightly yellow, and other times the central area is canary yellow, but the margins are usually white.

Semele purpurascens (Gmelin, 1791)   **Purplish Semele**

Distribution: North Carolina to Florida, Texas; Caribbean; Brazil.

Size: 25 to 32 mm (1 to 1¼ in).

Description: Color cream with maculations of purple or orange; inner surface with large area of rose, purple, or orange in umbonal cavity. Shape elongate-oval. Sculpture smooth except for fine commarginal growth lines. Umbones found posteriorly on shell. Anterior dorsal margin smoothly sloping to almost straight; posterior dorsal margin steep and convex. **Habitat:** Shallow, sandy bottoms along the Texas coast at depths from 1 to 55 m (3 to 180 ft). **Remarks:** Distinguishing characteristic is that the fine concentric threads intersect the commarginal growth lines and give a pattern of diagonally crossing threads.

Solecurtidae

Solecurtidae is a relatively small family of bivalves of about 30 species found in tropical and temperate marine and fresh waters. Shells are rectangular, long, and narrow. The upper and lower valves are almost equal, and both ends are rounded and gaping. Sculpture is smooth except for commarginal growth lines. The periostracum is well developed, and variation in color and thickness is seen within species. Beaks are relatively small. A strong external ligament is positioned behind the beaks on the ligamental shelf. Muscle scars are located toward the dorsal margin. The animal is quite large in the solecurtids, and the shell does not appear large enough to house the whole animal. The right valve has 1 or 2 small cardinal teeth. The left valve has 1 projecting tooth and weak lateral teeth. The pallial sinus is variable and may be deeply impressed or shallow. Most solecurtids live in clean substrate subtidally in permanent vertical burrows. The relatively deep burrows can be 500 mm (20 in) beneath the surface of the sediment. In Texas, Solecurtidae is represented by 2 genera and 3 species whose size ranges from 25 to 89 mm (1 to 3½ in); all 3 species are discussed here.

Solecurtus cumingianus (Dunker, 1861)

Corrugated Razor Clam

Distribution: North Carolina to Florida, Texas; Brazil.
Size: 25 to 76 mm (1 to 3 in).

Description: Color uniformly white. Shape elongate-cylindrical, somewhat inflated. Sculpture of commarginal growth lines diagonally crossed by wavy ridges most pronounced at center of exterior surface of valve. Shell gapes at both ends. Right valve with 2 distinct cardinal teeth; left valve with 1 cardinal tooth. **Habitat:** Offshore, infaunal. Typically found alive at depths of about 47 m (156 ft). Depth range from 8 to 183 m (26 to 600 ft). **Remarks:** Collected or dredged from the surface to a depth of about 126 m (420 ft). Rarely found alive.

Tagelus divisus (Spengler, 1794)   **Purplish Tagelus**

Distribution: Cape Cod, Massachusetts to Florida to Texas; Brazil; Bermuda.
Size: 25 to 38 mm (1 to 1½ in).

Description: Color somewhat translucent purplish-whitish, with major concentrations of purple toward the center of the valve. Shape elongate-cylindrical, compressed. Sculpture smoothish with faint, irregularly spaced commarginal growth lirae. Gaping anteriorly and posteriorly. Umbones suppressed; hinge area with 1 cardinal tooth in each valve; no lateral teeth. Adductor muscle scars double. Pallial sinus deep. **Habitat:** In muddy bottoms along the Texas coast at depths from 0 to 22 m (72 ft). **Remarks:** Common food source for herons at low tide.

Tagelus plebeius (Lightfoot, 1786)   **Stout Tagelus**

Distribution: Cape Cod, Massachusetts to Florida to Texas; Caribbean; Brazil.
Size: 51 to 89 mm (2 to 3½ in).

Description: Color uniformly white with a persistent brown periostracum. Shape elongate-rectangular, cylindrical. Sculpture smoothish with fine, irregularly spaced commarginal lirae. Shell gapes at both ends. Hinge area with 1 cardinal tooth in each valve; no lateral teeth. Adductor muscle scars double. Pallial line simple. **Habitat:** In bay margins among the roots of vegetation at depths from 0 to 6 m (20 ft). **Remarks:** Uses its strong foot to burrow into the mud vertically. Can tolerate lower saline conditions than _T. divisus._

Solenidae

Solenidae, commonly called the razor and jackknife clam family, is a relatively small-sized family of bivalves with 2 genera and approximately 60 species. These clams are found worldwide in shallow, warm tropical waters, usually in fine, clean sand. Shell shape and structure are cylindrical, narrow, and fragile; the shell gapes at both ends. Sculpture is smooth with the exception of commarginal growth lines. A smooth, glossy periostracum is present. Beaks are located terminally or subterminally (near extremity of shell). The hinge area is weak with 1 central tooth in each valve. Razor clams are equipped with a strong muscular foot that allows them to burrow rapidly into sandy substrate. Some species retain a permanent burrow, and a few can propel themselves along the bottom. In Texas, Solenidae is represented by 1 species.

Solen viridis (Say, 1822)   **Green Jackknife Clam**

Distribution: Rhode Island to Florida, Texas, Gulf of Mexico.

Size: 51 mm (2 in).

Description: Color semi-translucent white; periostracum glossy greenish to brownish. Shape long, almost 4 times as long as wide, somewhat flattened cylinder. Sculpture fragile, smooth; dorsal edge somewhat level; ventral edge slightly curved. Long, narrow, fine commarginal growth lines barely visible. Hinge plate with single prominent tooth at end of each valve. **Habitat:** Beach zone inlets; shallow-water sand flats. Found dead at depths from shoreline to 45 m (150 ft) and alive from shoreline to 12 m (40 ft). **Remarks:** Reaches its westernmost distribution in Texas. Easily confused with another jackknife clam, *Ensis minor,* but is not as narrow or as long.

Pharidae

Pharidae is a moderate-sized family of bivalves of about 65 species found worldwide. These elongate, slender bivalves are similar to Solenidae; however, the pharids are more compressed, the right valve has

2 cardinal teeth, and the left valve has1 cardinal tooth. Beaks are positioned subterminally. The shells are usually reinforced by calcareous rays originating inside the shell beneath the beaks. Sculpture is of commarginal growth lines only. The periostracum is similar to that of the Solenidae in that it is smooth and appears varnished. The pallial sinus is relatively deep and abrupt. Siphons are short to long and are either fused or separate. A small extra orifice, known as a fourth pallial aperture, is found in the genus *Ensis*. The foot is relatively large, obliquely flattened, and blunt at the end. In Texas, Pharidae is represented by 1 species.

Ensis minor (Dall, 1900)

Minor Jackknife Clam

Distribution: New Jersey to Florida to Texas.
Size: 76 mm (3 in).

Description: Color whitish; interior polished purplish-white. Shape long and slender, somewhat curved cylinder. Sculpture smoothish, fragile with irregularly spaced concentric, strong growth ridges with fine concentric growth lirae between; interior smooth with growth ridges showing through, tinged darker purple. Left valve with 2 cardinal teeth, 1 cardinal tooth on right valve, 1 lateral tooth in each valve. **Habitat:** Margins of bays and lagoons. Found dead at depths from shoreline to 101 m (330 ft) and alive from shoreline to 3 m (9 ft). **Remarks:** Common species along the Texas coast.

Mactridae

Mactridae, commonly known as the surf clam family or trough family, is a large family of bivalves of approximately 150 species. They are usually found in shallow marine habitats worldwide. Shell shape is typically oval to almost triangular and occasionally oblong. Sculpture is usually smooth but may have weak ridges. Beaks point forward and are from large and robust to small and acute. The inner ligament is triangular, sturdy, and located in a triangular depression or spoon-shaped appendage. The left valve is equipped with 1 V-shaped cardinal tooth located centrally; the right valve has 2 cardinal teeth. Side or lateral teeth are usually present and distinct. Muscle scars are present. The pallial apparatus is found ventrally and may be small but is well defined. Mactrids are equipped with 2 strong, short siphons and a moderate-sized, powerful foot capable of burrowing rapidly. Many of these bivalves are common and sometimes are the dominant organism in areas where they are found. In Texas, Mactridae is represented by 7 genera and 7 species whose size ranges from 8 to 178 mm (⅓ to 7 in); 7 species are discussed here.

Anatina anatina (Spengler, 1802)

Smooth Duck Clam

Distribution: North Carolina to Florida to Texas; Brazil.

Size: 51 to 76 mm (2 to 3 in).

Description: Color translucent whitish with mellow shades of yellow; interior coloration harmonious with exterior. Shape oval-elongate, thin, fragile, posterior end gaping. Sculpture smoothish with fine, irregular commarginal growth lines; sulcus extending from posterior-pointing umbo to posterior ventral angle. Short lateral tooth behind chondrophore in each valve. **Habitat:** Found alive on Gulf beaches at depths from low-tide line to 2 m (7 ft). Depth range from 0 to 134 m (440 ft). **Remarks:** Believed to be a surf-zone species.

Mactrotoma fragilis (Gmelin, 1791)   **Fragile Surf Clam**

Distribution: Massachusetts to Florida, Texas; Caribbean.
Size: 51 to 90 mm (2 to 3½ in).

Description: Color cream-white with a grayish periostracum; inner surface white. Shape elongate-oval. Sculpture smooth with fine commarginal growth lines. Umbones just forward of the middle of shell; 3 cardinal teeth in left valve; 2 small cardinal teeth in right valve; lateral tooth behind the chondrophore in both valves. **Habitat:** In Texas on sandy mud bottoms at depths from 27 to 45 m (90 to 150 ft). Depth range from 0 to 45 m (150 ft). **Remarks:** This is a common species along the Texas coast, being more common in the south. **Synonym:** _Mactra fragilis_ (Gmelin, 1791).

Mulinia lateralis (Say, 1822)   **Dwarf Surf Clam**

Distribution: Maine to Florida to Texas.
Size: 8 to 12 mm (⅓ to ½ in).

Description: Color uniformly white with a tannish-brown periostracum. Shape triangular. Sculpture smooth and glossy with fine commarginal growth lines; radial ridge extends from umbo to posterior ventral angle. Each valve with 2 cardinal teeth with a cone-shaped chondrophore; both sides with well-developed lateral teeth. **Habitat:** In sand in warm, shallow water. Depth range from 0 to 134 m (440 ft). **Remarks:** Found in different types of environments but usually collected alive in shallow bays.

Raeta plicatella (Lamarck, 1818) **Channeled Duck Clam**

Distribution: North Carolina to Florida, Texas; Caribbean to Argentina.

Size: 51 to 76 mm (2 to 3 in).

Description: Color uniformly dull whitish. Shape broad, elongate-ovate. Sculpture of regularly spaced, rounded, corrugated concentric ribs that show through interiorly with small, wrinkled, radial threads in inner spaces of outer concentric ribs. **Habitat:** Surf zone. Found alive at depths from shoreline to 2 m (7 ft). Depth range from 0 to 285 m (935 ft). **Remarks:** Dead shells are commonly found along the shoreline and dredged to a depth of 47 m (156 ft).

Rangia cuneata (Sowerby I, 1831) **Atlantic Rangia**

Distribution: North Chesapeake Bay to Texas.

Size: 25 to 80 mm (1 to 3 in).

Description: Color uniformly white; persistent brown periostracum. Shape ovate. Sculpture smoothish with fine commarginal growth lines; shell thick. Valves with two cardinal teeth that form an inverted V-shaped projection. Laterals serrated, almost extending to ventral margin. Prominent umbones located anteriorly. Anterior adductor muscle scar smaller than posterior adductor muscle scar; pallial sinus short but distinct. **Habitat:** Brackish-water species; river-influenced Texas bays and marshes. Shells found at depths from shoreline to 124 m (407 ft). **Remarks:** Accumulations of these shells are used as road material.

Rangianella flexuosa (Conrad, 1839)   **Brown Rangia**

Distribution: Louisiana, Texas; Veracruz, Mexico.
Size: 25 to 38 mm (1 to 1½ in).

Description: Color uniformly white; persistent brown periostracum. Shape elongate-oval to triangular. Sculpture smoothish with fine commarginal growth lines; shell strong, thick; posterior sulcus extends from umbo to posterior ventral angle. Well-developed chondrophore; posterior lateral tooth extends to about midpoint between umbo and ventral angle.
Habitat: Brackish water. Shells found at depths from 0 to 84 m (275 ft).
Remarks: _Rangianella flexuosa_ has a similar habitat but cannot tolerate lower salinity levels, as can _Rangia cuneata_.

Spisula raveneli (Conrad, 1831)   **Southern Surf Clam**

Distribution: Nova Scotia to South Carolina, Texas.
Size: 178 mm (7 in).

Description: Color yellowish-white with light brown periostracum; interior white. Shape oval-elongate. Sculpture smoothish with fine, irregularly spaced commarginal growth lines. Umbones somewhat pointed, located anteriorly; well-developed chondrophore; posterior and anterior laterals narrow and furrowed. Pallial apparatus distinct; circular adductor muscle scars located above midpoint of shell. **Habitat:** Alive in surf zone and sandy substrate at depths from 4 to 22 m (12 to 72 ft). Shells found at depths from 0 to 90 m (300 ft). **Remarks:** Located all along the Texas coast and into Mexico.

Dreissenidae

Dreissenidae is a small family of bivalves of approximately 40 species found in brackish and fresh water. Shells are long and triangular and resemble those of the Mytilidae. Sculpture is typically smooth except for commarginal growth lines. Beaks are positioned at the acutely tapered anterior end. They all possess muscle scars. The anterior muscle is on the inner portion of the shell behind the beaks on a triangular platform. The posterior muscle is distinctly elongate on the opposite side. The animal is equipped with 2 abrupt siphons and a byssus. The dreissenids use this byssus to attach themselves to hard substratum. In Texas, Dreissenidae is represented by 1 species.

Mytilopsis leucophaeata (Conrad, 1831) **Dark False Mussel**

Distribution: New York to Florida to Texas; Mexico.
Size: 12 to 18 mm (½ to ¾ in).

Description: Color whitish with persistent brown, fibrous periostracum; inner surface white, stained with bluish-gray. Shape elongate-triangular, mussel-like. Sculpture absent except for coarse, concentric growth lines. Hinge area of margins almost parallel; ventral margin well rounded. Umbones at narrow pointed end; long byssal gape; triangular element located beneath apical shelf. **Habitat:** Attached to hard substrates at surface in brackish to freshwater environments. Alive from surface to a depth of 2 m (7 ft). Depth range from 0 to 55 m (180 ft). **Remarks:** Wide occurrence in brackish to fresh water from Mississippi Delta to Rio Grande.

Myidae

Myidae is a small family of clams of approximately 20 species located in all seas; however, they are best represented in the northern hemisphere. Shells are fragile, widely oval to irregularly oblong with a posterior gape, glossy, often whitish with smooth sculpture or irregular commarginal growth lines with a thin periostracum. The left valve has a platform for the internal ligament. The right valve has a pit under the beak to accommodate the corresponding ligament from the left valve. A pallial sinus is sometimes present; the pallial line is undulating and interrupted. Muscle scars are almost equal. Myids typically have 1 long, fused siphon. Some myids are able to go for days without oxygen. Myids survive by either boring into mud or sand or huddling in rocky crevices or old bore holes. In Texas, Myidae is represented by 2 genera and 2 species whose size ranges from 5 to 18 mm (⅕ to ¾ in); 1 species is discussed here.

Paramya subovata (Conrad, 1845)

Subovate Softshell

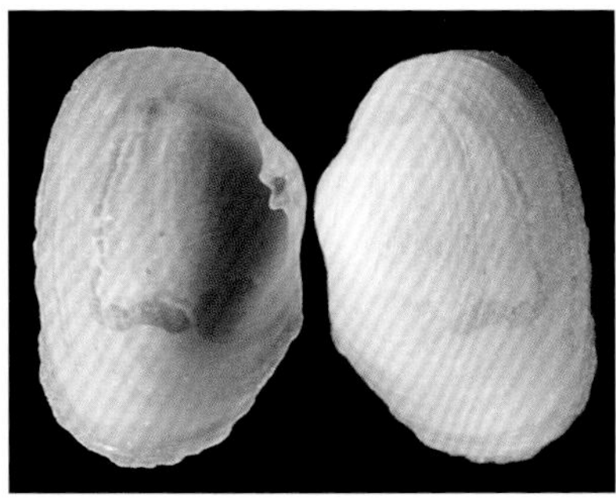

Distribution: Delaware to Florida, Texas.
Size: 5 to 10 mm (⅕ to ⅖ in).

Description: Color uniformly grayish-white. Shape oblong-subquadrate. Sculpture of fine commarginal growth lines with exterior margins possessing microscopic granulations. Pallial line obscure; large rounded adductor muscle scar located dorsally. **Habitat:** Commensal relationship with the euchrid worm (_Thalassema hartmani_) in offshore waters. Depth range from 0 to 46 m (150 ft). **Remarks:** A study by Henry (1976) found over 200 living specimens of _P. subovata_ associated with numerous _T. hartmani._

Corbulidae

Corbulidae, the largest family in the order Myoidea, is a moderate-sized family of bivalves of about 75 species found in all seas. Shells can be elongate, widely ovate, to somewhat triangular with an acutely tapered posterior end. Sculpture is often concentrically ribbed but sometimes semi-smooth. The corbulids are easily distinguished from other Myoidea because of their unusual shape; they have unequal valves and have a channel-like apparatus that extends from the beaks to the posterior ventral angle. The shell is thick and typically convex. The right valve is larger than the left valve, and the lower margin of the right valve overlaps the lower margin of the left. A tooth in the right valve fits into a notch under the beak of the left valve. The pallial sinus is reduced or absent. Corbulids live buried in muddy sand or gravel or attached by byssal threads to hard substratum. Valves of young Corbulidae are thin but quickly become thick with age, which may cause taxonomic confusion. In Texas, Corbulidae is represented by 2 genera and 5 species whose size ranges from 6 to 12 mm (¼ to ½ in); 2 species are discussed here.

Caryocorbula caribaea (d'Orbigny, 1853) — Caribbean Corbula

Distribution: Massachusetts to Florida, Texas; Caribbean.
Size: 6 mm (¼ in).

Description: Color uniformly dull white. Shape irregular, elongate, slightly inflated. Sculpture smoothish with irregularly spaced, fine commarginal growth lines. Right valve overlaps left valve at posterior, ventral margin. Gentle ridge extends from umbones to posterior ventral angle. **Habitat:** Inlets, bays, and beach drift at depths to 137 m (450 ft). **Synonym:** *Corbula swiftiana* (C. B. Adams, 1852).

Varicorbula limatula (Conrad, 1846) — Oval Varicorbula

Distribution: North Carolina to Florida, Texas; Caribbean, Brazil.
Size: 9 mm (3/8 in).

Description: Typically whitish, sometimes with a rose color around the margins. Shape subcircular to ovate. Sculpture of irregular, wrinkled growth lines. Umbones elevated, curved inward, and point anteriorly. Left valve smaller than overlapping right valve. Obscure broad pallial sinus seen in fresh shells. **Habitat:** Offshore on continental shelf. Shells found at depths from 0 to 549 m (1800 ft). **Remarks:** Commonly collected dead on the beach from St. Joseph Island southward. **Synonyms:** *V. operculata* (Philippi, 1848); *V. disparilis* (d'Orbigny, 1853).

Pholadidae

Pholadidae is a moderate-sized family of bivalves of about 92 species found in all tropical and temperate waters. Shells are irregular, thin, elongate, oblong, ovate to almost circular, and widely gaping. Posteriorly there is an almost circular gape. In some species the gape may be partly or totally closed by calcareous plates. Sculpture is typically of thin, concentrically layered ridges and riblets. The irregular hinge plate is equipped with articulating processes for attachment of muscle. An obscure furrow that extends from the umbones to the posterior end typically separates sections of the shell and produces different sculpture on each section. The pallial sinus is wide, and the pallial line is impressed. These bivalves are highly specialized for burrowing and boring into firm or hard substrate and are believed to use chemical secretions in some instances. The pholadids bore into semi-hard clay, mud, soft rock, or even plastic cables, which can be destructive. In Texas, Pholadidae is represented by 6 genera and 8 species whose size ranges from 12 to 178 mm (½ to 7 in); all 8 species are discussed here.

Barnea truncata (Say, 1822)

Fallen Angelwing

Distribution: Massachusetts to Texas; Brazil.

Size: 51 to 70 mm (2 to 2¾ in).

Description: Color uniformly dull, chalky white. Shape elongate-rectangular, inflated, gaping at both ends. Sculpture of strong concentric ribs anteriorly, becoming less distinct posteriorly; radial ribs strongest on anterior slope, becoming weaker at central portion and obscure to absent posteriorly. Umbones positioned anteriorly. Pallial apparatus well defined. Apophysis (bony spoon-shaped protuberance) elongate, slender, and blade-like. **Habitat:** Burrows into mud, peat, and clay. In Texas found alive burrowing in mudflats off Bolivar Peninsula at depths from shoreline to about 2 m (7 ft).

Cyrtopleura costata (Linnaeus, 1758)

Distribution: Massachusetts to Texas; Brazil.
Size: 102 to 178 mm (4 to 7 in).

Description: Color uniformly white. Shape elongate-oval, cylindrical. Sculpture of strong radial ribs almost covering the entire length of the shell, with concentric ribs producing spinelike projections where they cross; most prominent spines located anteriorly. Umbones somewhat separated from umbonal reflections; protoplax (supplementary plate in front of the umbone) triangular, chitinous; mesoplax (shelly piece located above the umbone) calcareous; apophysis spoon shaped with upper portion hollow. **Habitat:** Burrows into bay margins and inlets; infaunal. Angelwings are found alive in South Texas bays at depths from shoreline to about 2 m (7 ft). Depth range from 0 to 49 m (160 ft). **Remarks:** Commonly washed ashore in beach drift.

Diplothyra smithii (Tryon, 1862) **Oyster Piddock**

Distribution: Massachusetts to Texas.

Size: 12 mm (½ in).

Description: Color white to light brown; inner surface somewhat polished. Shape oblong-oval, pear shaped. Sculpture of fine, undulating concentric threads crossed by almost indistinguishable radial lirae at anterior end; posterior end compressed with sculpture of only commarginal growth lines. Pedal gape found posteriorly and closed by a callus, which leaves only a slitlike opening between the 2 valves. Apophysis long and narrow. **Habitat:** Bores into oyster shells on oyster reefs; infaunal. Depth range from 0 to 42 m (140 ft). **Remarks:** One of the most common species of boring bivalves in Texas bays; typically found in old oyster shells but sometimes found in concrete and soft rock.

Jouannetia quillingi (Turner, 1955) **Spiny Piddock**

Distribution: North Carolina to Florida to Texas.

Size: Up to 21.5 mm (¾ in).

Description: Color uniformly chalky white. Shape suborbicular, fragile. Sculpture differentiated into 2 parts by umbonal sulcus; well-developed callus; right valve with concentric ribs at angle to sulcus and somewhat obscure radial ribs; left valve sculptured of irregular, thin concentric threads with horizontal inference of concentric threads at sulcus. Gaping anteriorly; closed and spherical posteriorly. **Habitat:** Bores into hard substrates offshore; infaunal at depths from 0 to 68 m (223 ft). **Remarks:** Quite common at Stetson Bank and other shale uplifts of the Texas coast.

Martesia cuneiformis (Say, 1822)   **Wedge Piddock**

Distribution: North Carolina to Texas to Brazil.
Size: 12 to 18 mm (½ to ¾ in).

Description: Color uniformly white. Shape pearlike. Anterior sculpture of closely spaced, concentric crenulate ribs; posterior sculpture with gently rounded concentric ribs and fine commarginal growth lines. Umbones distinct; mesoplax wedge shaped; hypoplax (supplemental plate that covers the gape in the postero-ventral region) and metaplax elongate, slender, and separated at the hind end, producing a callus in adulthood. **Habitat:** Bores into driftwood; pelagic; infaunal at depths from 0 to 37 m (121 ft). **Remarks:** Easily confused with *Diplothyra smithii* unless the cuneiform or wedge-shaped mesoplax is present.

Martesia fragilis (Verrill and Bush, 1898)   **Fragile Piddock**

Distribution: Virginia to Texas; Brazil; Bermuda.
Size: 12 to 18 mm (½ to ¾ in).

Description: Color uniformly white. Shape somewhat pearlike. Sculpture of valves divided into 2 distinct parts by a shallow sulcus; anterior portion with concentric, wavy, crenulate ribs; posterior end with gently rounded ridges. Umbones distinct, located anteriorly; mesoplax round, depressed with strong concentric sculpture; hypoplax and metaplax elongate, slender, and separated at the hind end, producing a callus in adulthood. **Habitat:** Bores into driftwood; pelagic; infaunal at depths from 0 to 51 m (167 ft). **Remarks:** Easily confused with *M. striata;* however, the mesoplax is rounded and depressed.

Martesia striata (Linnaeus, 1758)   **Striate Piddock**

Distribution: North Carolina to Texas to Brazil.
Size: 18 to 45 mm (¾ to 1¾ in).

Description: Color uniformly white. Shape somewhat pearlike, variable. Anterior sculpture of closely spaced, concentric, crenulate ribs; posterior sculptured with gently rounded, concentric ribs. Umbones distinct, located anteriorly in adult shells with a sickle-type flange over the umbones; metaplax and hypoplax elongate, slender, and acute at both ends. **Habitat:** Bores into driftwood; pelagic; infaunal at depths from 0 to 21 m (70 ft). **Remarks:** Easily confused with _M. fragilis;_ however, the mesoplax is not depressed, and the shell appears more elongate and slender.

Pholas campechiensis (Gmelin, 1791)   **Campeche Angelwing**

Distribution: North Carolina to Texas to Brazil.
Size: 76 to 114 mm (3 to 4½ in).

Description: Color uniformly chalky to grayish-white. Shape long, oval-cylindrical, fragile with a slight gape at both ends. Sculpture of concentric growth lines and narrow radial ribs, becoming somewhat scaly at intersection, most pronounced anteriorly. Umbones positioned anteriorly, covered by a reflecting plate; long apophysis curves out from beneath umbones. **Habitat:** Burrowed in mud below low-tide mark to a depth of 9 m (30 ft). **Remarks:** Dead shells are somewhat common in beach drift but have been collected alive offshore McFadden Beach near Sabine Pass.

PART V

CLASS: SCAPHOPODA
Tuskshells

Scaphopoda is a relatively small class of burrowing mollusks of approximately 350 species, commonly known as toothshells or tuskshells. Scaphopods are cosmopolitan benthic carnivores. The tubular shell is sturdy, long, curved, tapered, and open at both ends. In their natural habitat they lie with the wider end, or aperture, housing the head and foot, submerged in sediments and with the smaller, tapered end, or apex, projecting out of the substrate. Water and particles are drawn in, and waste products are removed through the apex. The foot is strong and adapted for digging. The head lacks eyes and tentacles. The mouth is equipped with a radula surrounded by threadlike projections used for capturing prey. Scaphopods primarily feed upon foraminifera and tiny bivalves. The apex may have slits or notches in it. Sculpture may be smooth with longitudinal ridges or with spiral rings. Gills are absent. Scaphopods take in oxygen by absorption through the mantle. They are sexually dimorphic; sperm and eggs are siphoned out through the apex into the water column. After fertilization the eggs hatch, and the free-swimming larvae settle on the bottom after 5 to 6 days. In Texas, 4 families and 10 species are present; the most common species is discussed here.

Dentaliidae

Dentaliidae is a family of scaphopods made up of approximately 14 genera and about 240 species worldwide. They are found in all waters from shallow to abyss. Members of the family Dentaliidae are almost straight to distinctly curved, hardy shells. Sculpture is longitudinal and varies from a few distinct robust ribs to many fine threads that are sometimes limited to the apical part of the shell. The apical portion of the shell shows a variety of forms, from plain to having a notch, slit, or series of minute holes. The posterior opening may sometimes show different characters, such as having a plug that narrows the opening or a short tube attached to it. Dentaliids typically exhibit a white coloration, but some can have greenish, yellowish, or pink pigmentation. The greatest diversity among the Dentaliidae occurs in tropical and subtropical waters at depths from the upper subtidal zone to approximately 1000 m (3330 ft). The genus *Dentalium* is the most speciose of all scaphopods. In Texas, Dentaliidae is represented by 5 genera and 6 species whose size ranges from 18 to 100 mm (¾ to 4 in); only the most common species is discussed here.

Paradentalium americanum (Chenu, 1843) **Texas Tuskshell**

Distribution: North Carolina to Texas.

Size: 18 to 55 mm (¾ to 2⅕ in).

Description: Color dull white. Shape well arched. Sculpture of 6 distinct ribs, giving a hexagonal structure; inner spaces between flattened longitudinal ribs with numerous annular riblets. Aperture hexagonal. **Habitat:** Clay sediments. Semi-infaunal. Found in open-bay margins and inlets at depths from 1 to 95 m (3 to 312 ft).
Synonyms: *Dentalium americanum* Chenu, 1843; *D. texasianum* (Philippi, 1849).

adductor muscle scars. Depressions that mark the attachment area of the adductor muscles.

adpressed. Lying pressed close to one another; relating to overlapping whorls with their surfaces gradually converging.

anal canal. Found in dorsal area of aperture in gastropods and as an excurrent siphon in bivalves; area where rectum empties.

apertural lip. Outer margin of the aperture in gastropods.

aperture. Open portion of a gastropod shell; where the head and foot appear.

apex. Point at the tip of the spire of a gastropod; the extreme top of the spire and usually consists of the protoconch.

apophysis. A bony, spoon-shaped, calcareous protuberance found in some mollusks (e.g., Pholadidae).

articulate. Consisting of sections united by joints, such as the interlocking teeth of the hinge plate in bivalves.

axial. Along longitudinal but not spiral axis; parallel to the axis of coiling in gastropods.

base. Region of gastropod opposite the apex of the spire; basal: the bottom or lower part.

beak. Pointed extremity of a bivalve shell at which it began to grow; the apex or umbo of a bivalve shell; the beak is either straight as in *Argopecten,* curved as in *Mercenaria,* or spiral as in *Glossus humanus.*

benthos. The whole assemblage of organisms living in or upon the sea bottom.

body whorl. The last and usually the largest whorl of a gastropod shell.

byssus. The beard; small bundle of silky threads secreted by the bivalve foot by which mollusks attach themselves to hard surfaces.

callus. An unusually hardened part of the shell, usually external, secreted by the mantle.

canal. A groove or duct, such as the anterior siphonal canal in *Murex*.

cancellate. Having spiral ribs crossed by axial ribs of equal development; latticed; having a network formed by interlacing bars (e.g., *Chione elevata*).

cardinal teeth. Central, primary teeth of the hinge.

carina (plural, carinae). A keel-like projection or bladelike ridge on the outer surface of a shell.

chink. Narrow cleft, crack, or slit, as on the margin of the columella in *Probythinella protera* (e.g., umbilical chink).

chitinous. Referring to a colorless, hard, complex carbohydrate that forms the exoskeleton of insects and crustaceans, the ligament of bivalves, the internal shell of squids, and the horny operculum of many gastropods, as *Littoraria* and *Architectonica*.

chondrophore. Spoon-shaped object projecting from the hinge plate.

columella. The mostly hidden axial or central pillar of gastropod shells around which the whorls are built; extends from apex to base; a portion of the columella can be seen at the aperture of most spiral gastropods.

commarginal. Having ridges, ribs, or growth lines that run parallel to the margin of the shell; also called *concentric*.

compressed. Laterally or dorso-ventrally flattened as in some bivalves; nearly flat; reduced thickness.

concentric. Having circular lines or ridges curving about a middle area; applied to curved ridges on a bivalve with arcs having the same center; opposed to radial lines or ridges; also called *commarginal*.

conchology. Scientific study that describes mollusks based upon shell characters or hard parts of the shell.

costa (plural, costae). Tubular or circular ridge on the surface of the shell, larger than a cord.

crenulate. Having a series of notches or corrugated wrinkles on the margins of shells; having indented or scalloped shell margin; regularly indented.

deck. Small, thin plate of shell-like material found in the umbonal region connecting the anterior and posterior ends of a valve; also called a *septum* and used to describe the shell-like covering of the aperture in *Crepidula*.

dentate. Having processes resembling toothlike structures, as in the aperture of some gastropods and hinge of some bivalves.

denticles. Minute projecting points of toothlike structures.

dextral. Right-handedness in mollusks; opposed to *sinistral*.

divaricate. To extend in opposite directions from a common point; spread apart; branch off.

dorso-ventral. Relating to both the dorsal and ventral portions of the body.

duplivincular. Referring to a type of dorsal ligament made up of a series of bands attached to a shallow groove in the cardinal area of some bivalves (e.g., Arcidae, Noetiidae).

edentate. Lacking teeth or folds, as on the hinge plate of some bivalves and the outer lip or columella of the aperture in some gastropods.

epifauna. Organisms that live "on" top of the benthos (epibenthic) (as opposed to infauna, which live "in" the bottom).

equilateral. In bivalves, having the right and left valves located centrally and of equal size.

equivalve. Having the 2 valves in the same shape and of equal size.

escutcheon. A heart-shaped depression on the postero-dorsal surface of a bivalve shell, usually demarcated by a marginal furrow; may be on one or both valves.

fissure. A narrow slit or opening; crevice; furrow.

flammule. Maculation of color similar in appearance to a small flame.

fold. A spiral ridge on the columella extending into the interior of the shell.

foot. Muscular, fleshy organ of locomotion found on the undersurface of the body of a mollusk, upon which the animal rests or moves.

fusiform. Torpedo shaped; spindle shaped with a long canal and an equally long spire, tapering from the middle toward each end; applied to gastropods.

gape. A natural opening found along the margin of bivalve shells that do not shut tight, as in soft-shelled clams; gaper (e.g., *Solen, Ensis*).

hectocotylus. Modified arm of the male cephalopod that serves as an instrument of copulation.

hermaphrodite. An individual with both male and female reproductive organs.

hinge. Interlocking toothed mechanism in a bivalve.

hypoplax. A supplemental plate that covers the gape in the postero-ventral region of some Pholadidae.

incised. Grooved; channeled; sculptured with sharply cut grooves.

infauna. Organisms living "in" the bottom of the sea (as opposed to epifauna, which live "on" the bottom).

keel. Longitudinal cord; carina; a distinct spiral ridge that usually indicates a change of slope in the outline of the shell.

ligament. An elastic band, usually black or brown, that connects and opens the valves of bivalves when adductor muscles relax; located above the hinge, usually posterior to the umbones; usually the greater part of the ligament is externally placed but may be entirely or partially internal in some genera.

lira (plural, lirae). Fine raised line or groove on the surface of a shell.

lithodesma. A calcareous strengthening of the internal ligament.

longitudinal. Running the length of the shell; referring to the direction of the longest diameter (e.g., longitudinal ribs).

lunule. Crescent- or heart-shaped impression in front of the umbones in a bivalve, one-half being on each valve; a crescent-shaped part or marking (e.g., *Mercenaria campechiensis*).

malacology. The scientific study of mollusks; the branch of zoology that deals with mollusks, both the soft parts and the shell.

mantle. The membranous covering of a mollusk that secretes the shell from marginal glands and produces the periostracum; pallium.

margin. The outer edge of a shell.

mesoplax. A calcareous shelly piece located above the umbone of some pholads.

metaplax. Accessory plate behind the umbone of some pholads.

mucro. The pointed, posterior septum of *Caecum* shells; also known as *apical plug;* the high end of the tail valve of chitons.

muscle scar. A spot or depression inside a bivalve shell that indicates where the adductors were attached.

nacreous. Having an iridescent luster, as in oyster shells.

nautiloid. Resembling a *Nautilus* in shape; referring to the shape of any cephalopod shell spiraled in a symmetrical involute coil, as in the genus *Nautilus*.

nodulose. Having knobs of small size.

notch. A V-shaped indentation usually found on an edge or surface of a shell; a nick, indentation, or irregularity in the peristome that denotes the position of the siphon.

nucleus. Earliest part of a shell; formed when the larva hatches from the egg; embryonic shell that remains and becomes the apex of the adult gastropod shell.

operculum. A plate that plugs the aperture of a gastropod either wholly or partly when the snail is retracted; primarily for protection. Most gastropods have a corneous operculum, but it can be calcareous as in Turbinidae.

ornamentation. Sculpture found on mollusks.

outer lip. Outer margin of the aperture from the suture to the base of the columella.

ovate-subquadrate. Oval but somewhat squared off.

pallial line. A single-lined impression in a bivalve produced by the margin of the mantle.

pallial sinus. A cut or indentation in the pallial line.

parietal. Relating to the inner wall of the aperture of a gastropod; relating to the wide upper part of the inner lip.

patelliform. Having the shape of a flattened cone (e.g., Patellidae and Fissurellidae).

pen. An inner shell found in most squids; may be slender, very thin, delicate, horny, or lanceolate.

penultimate whorl. The whorl immediately following the body whorl; next-to-last whorl.

periostracum. A protective skin or horny covering that encloses the shell and protects against erosion; sometimes fragile and translucent.

plait. Fold found on the columella of gastropods; there may be multiple columellar folds.

planispiral. Referring to the shape of a shell that winds in one plane, like a flat spiral.

predaceous. Carnivorous; living by preying upon other animals.

proboscis. An elongated appendage derived from the mouthparts of some invertebrates.

prodissoconch. The embryonic shell of a bivalve; equivalent to the protoconch of gastropods.

protoconch. Embryonic shell or nuclear whorls of a shell of a gastropod; frequently different in design, texture, or color from the adult shell and often important for identification.

protoplax. A supplementary plate in front of the umbone in some pholads (e.g., *Cyrtopleura costata*).

punctate. Covered or studded with numerous tiny holes over the surface.

pustulate. Like a bubble or blister; describing a growth resembling a pimple, blister, or wartlike projection that forms pustules.

radial symmetry. Having similar parts regularly arranged around a central axis.

radula. A rasping organ or tooth; a tonguelike ribbon armed with teeth and used extensively in taxonomy; found in nearly all mollusks except bivalves.

ray. One of a number of fine lines radiating from a center.

resilifer. Depression on the hinge plate supporting the internal ligament or resilium.

resilium. The internal ligament in bivalves, ventral to the hinge line.

reticulate. Having a cancellate sculpture; cross-ridged; having distinct ribs crossing each other like a network.

rhinophores. A pair of secondary earlike tentacles on the head of nudibranchs.

rib. A long and narrow slightly elevated strip; an elongated ridge.

scales. Usually pertains to the overlapping or closely set calcareous structures on the dorsal and/or exterior side of mollusks.

shouldered. Referring to an area adjacent to or along the edge of a higher more prominent part, as the whorls in some gastropods.

sinistral. Left-handed; with whorls of a spiral shell turning toward the left; winding to the left when the spire is upward; counterclockwise; opposed to *dextral.*

siphonal canal. A channeled extension of the aperture at the enclosure of the siphon.

siphonal notch. A narrow cut or slit at the margin of the aperture near the base of the columella; hallmark of the longer, more recently evolved siphonal canal.

spine. Sharp pointed structure or outgrowth.

spire. A coil; upper series of whorls of a gastropod except the last one.

stenotypic. Having narrow ranges of tolerance; confined; referring to species in a community that are restricted to a particular area; opposite *eurytypic.*

stromboid notch. Referring to a notch in the outer lip of a *Strombus,* located just above the anterior notch.

sulcus. A groove, longitudinal ridge, or channel.

suture. The line of junction or seam along which 2 hard structures join; a continuous spiral line marking the junction of whorls in a gastropod shell.

symmetry. Balanced, suitable arrangement; equal sided; with common parts arranged in mirror image along an axis.

taxodont dentition. A series of small parallel to subparallel teeth that are perpendicular to the hinge line, as in the family Arcidae.

teeth. Pointed protuberances at the hinge of bivalve shells that come together and interlock with corresponding sockets of opposite valves; the toothlike structures in the aperture of some gastropods, as on the inner lip of *Nerita* or on the outer lip of *Cassis*.

teleoconch. Excluding the protoconch, the entire gastropod shell.

trochiform. Top shaped; having a flat-sided conical shell (e.g., Trochidae).

turbinate. Top shaped; shaped like a top or an inverted cone; shaped like a reversed cone with a wide base.

turrid notch. The anal notch on the aperture of gastropods in the family Turridae.

umbilicus. A sunken area or depression at the axial base of a gastropod; "navel."

umbo (or umbone; plural, umbones). Beak; point of a bivalve found just above the hinge; the first formed part of a bivalve.

varix. One of the longitudinal thickened protuberances interspersed on the outer surface or across the whorls of some gastropods; usually suggests a process that occurs at regular intervals in a resting stage; a former position of the outer lip of the aperture.

whorl. A spiral or turn of a gastropod shell; the largest whorl is the body whorl or the last whorl of a gastropod.

width. The greatest dimension measured at right angles to the length or height of a shell.

wing. A projection, expansion, or earlike extension at the hinge line of a bivalve.

BIBLIOGRAPHY

Abbott, R. T. 1974. American seashells. 2nd ed. New York: Van Nostrand Reinhold.

Andrews, J. 1977. Shells and shores of Texas. Austin: University of Texas Press.

Barrera, N. C. 2001. Micromolluscan assemblages on the Flower Garden Banks, northwestern Gulf of Mexico. Master's thesis, Department of Physical and Life Sciences, Texas A&M University–Corpus Christi.

Beesley, P. L., G. J. B. Ross, and A. Wells, eds. 1998. Mollusca: The southern synthesis. Fauna of Australia. Vol. 5. Melbourne, Australia: CSIRO Publishing.

Blankinship, R., M. Lingo, P. Trial, and R. Bennett. 2005. A study of the live mollusk and echinoderm harvest of the Lower Laguna Madre, Texas. Management Data Series No. 240. Austin: Texas Parks and Wildlife Department, Coastal Fisheries Division.

Britton, J. C. 1970. The Lucinidae (Mollusca: Bivalvia) of the western Atlantic Ocean. PhD diss., George Washington University, Washington, DC.

Britton, J. C., and B. Morton. 1989. Shore ecology of the Gulf of Mexico. Austin: University of Texas Press.

Brusca, R. C., and G. J. Brusca. 2003. Invertebrates. 2nd ed. Sunderland, MA: Sinauer Associates.

Collin, R. 2002. Another last word on *Crepidula convexa* with a description of *C. ustulatulina* n. sp. (Gastropoda: Calyptraeidae) from the Gulf of Mexico and southern Florida. Bulletin of Marine Science 70:177–184.

Felder, D. L., and D. K. Camp, eds. 2009. Gulf of Mexico—origin, waters, and biota. Vol. 1. Biodiversity. College Station: Texas A&M University Press.

Fretter, V., and M. A. Graham. 1962. British prosobranch molluscs. Their functional anatomy and ecology. London: Ray Society.

Geiger, D. L., B. A. Marshal, W. F. Ponder, T. Sasaki, and A. Warén. 2007.

Techniques for collecting, handling, preparing, storing and examining small molluscan specimens. Molluscan Research 27(1):1–50.

Hicks, D. W. 2010. Molluscan ecology and habitats. Pp. 28–75 in J. W. Tunnell Jr., J. Andrews, N. C. Barrera, and F. Moretzsohn, Encyclopedia of Texas seashells: Identification, ecology, distribution, and history. College Station: Texas A&M University Press.

Lee, H. G. 2009. Marine shells of northeast Florida. Jacksonville Beach, FL: Jacksonville Shell Club.

Mikkelsen, P. M., and R. Bieler. 2008. Seashells of southern Florida. Living marine mollusks of the Florida Keys and adjacent regions. Bivalves. Princeton, NJ: Princeton University Press.

Moore, D. R. 1972. Ecological and systematic notes on Caecidae from St. Croix, U.S. Virgin Islands. Bulletin of Marine Science 22:881–899.

Odé, H. 1986. Distribution and records of the marine Mollusca in the northwest Gulf of Mexico (a continuing monograph). Texas Conchologist 23(1):20–36.

Odé, H. 1989. Distribution and records of the marine Mollusca in the northwest Gulf of Mexico (a continuing monograph). Texas Conchologist 25:59–80.

Redfern, C. 2013. Bahamian seashells: 1161 species from Abaco, Bahamas. Boca Raton, FL: Bahamianseashells.com.

Reid, D. G. 2009. The genus *Echinolittorina* Habe, 1956 (Gastropoda: Littorinidae) in the western Atlantic Ocean. Zootaxa 2184:1–103.

Roper, C. F. E., and E. K. Shea. 2013. Unanswered questions about the giant squid *Architeuthis* (Architeuthidae) illustrate our incomplete knowledge of coleoid cephalopods. American Malacological Bulletin 31(1):109–122.

Sturm, C. F., T. A. Pearce, and A. Valdés, eds. 2006. The mollusks: A guide to their study, collection, and preservation. Boca Raton, FL: American Malacological Society.

TPWD (Texas Parks and Wildlife Department). 2006. Outdoor annual hunting and fishing regulations 2006–2007. Austin: Texas Monthly Custom Publishing.

Tunnell, J. W., Jr., J. Andrews, N. C. Barrera, and F. Moretzsohn. 2010. Encyclopedia of Texas seashells: Identification, ecology, distribution, and history. College Station: Texas A&M University Press.

Turgeon, D. D., J. F. Quinn Jr., A. E. Bogan, E. V. Coan, F. G. Hochberg, W. G. Lyons, P. M. Mikkelsen, R. J. Neves, C. F. E. Roper, G. Rosenberg, B. Roth, A. Scheltema, F. G. Thompson, M. Vecchione, and J. D. Williams. 1998. Common and scientific names of aquatic invertebrates from the United States and Canada: Mollusks. 2nd ed. American Fisheries Society Special Publication 26. Bethesda, MD: American Fisheries Society.

Valentich-Scott, P., and E. Coan. 2012. Bivalve seashells of tropical west America: Marine bivalve mollusks from Baja California to Peru. Santa Barbara Museum of Natural History Monographs 6, Studies in Biodiversity 4. Santa Barbara, CA: Santa Barbara Museum of Natural History.

Vecchione, M. 2002. Cephalopods. Pp. 149–244 in K. E. Carpenter, ed., The living marine resources of the western central Atlantic. Vol. 1. Introduction, mollusks, crustaceans, hagfishes, sharks, batoid fishes and chimaeras. FAO Identification Guide for Fisheries Purposes. Rome: Food and Agriculture Organization of the United Nations.

Yonge, C. M., and T. E. Thompson. 1976. Living marine molluscs. London: Collins.